Homology and Systematics

T0179243

The Systematics Association Special Volume Series

Series Editor

Alan Warren
*Department of Zoology, The Natural History Museum,
Cromwell Road, London, SW7 5BD, UK*

The Systematics Association provides a forum for discussing systematic problems and integrating new information from genetics, ecology and other specific fields into taxonomic concepts and activities. It has achieved great success since the Association was founded in 1937 by promoting major meetings covering all areas of biology and palaeontology, supporting systematic research and training courses through the award of grants, production of a membership newsletter and publication of review volumes by its publishers Taylor and Francis. Its membership is open to both amateurs and professional scientists in all branches of biology who are entitled to purchase its volumes at a discounted price.

The first of the Systematics Association's publications, *The New Systematics*, edited by its then president Sir Julian Huxley, was a classic work. Over 50 volumes have now been published in the Association's 'Special Volume' series often in rapidly expanding areas of science where a modern synthesis is required. Its *modus operandi* is to encourage leading exponents to organise a symposium with a view to publishing a multi-authored volume in its series based upon the meeting. The Association also publishes volumes that are not linked to meetings in its 'Volume' series.

Anyone wishing to know more about the Systematics Association and its volume series are invited to contact the series editor.

Published titles in the series:

Molecular Systematics and Plant Evolution
Edited by P. M. Hollingsworth, R. M. Bateman and R. J. Gornall

Other Systematics Association publications are listed after the index for this volume.

The Systematics Association Special Volume Series 58

Homology and Systematics

Coding characters for phylogenetic analysis

Edited by Robert Scotland and R. Toby Pennington

CRC Press
Taylor & Francis Group
Boca Raton London New York

CRC Press is an imprint of the
Taylor & Francis Group, an **informa** business

CRC Press
Taylor & Francis Group
6000 Broken Sound Parkway NW, Suite 300
Boca Raton, FL 33487-2742

First issued in paperback 2019

© 2000 Systematics Association
CRC Press is an imprint of Taylor & Francis Group, an Informa business

No claim to original U.S. Government works

ISBN-13: 978-0-7484-0920-4 (hbk)
ISBN-13: 978-0-367-39889-7 (pbk)

British Library Cataloguing in Publication Data
A catalogue record for this book is available from the British Library

Library of Congress Cataloging in Publication Data
Scotland, Robert W.
   Homology and systematics : coding characters for phylogenetic analysis / Robert Scotland, Toby Pennington.
      p. cm. – (Systematics Association special volume ; no. 58)
   Includes bibliographical references.
   ISBN 0-7484-0920-3
      1. Homology (Biology). 2. Cladistic analysis–Methodology. I. Pennington, Toby. II. Systematics Association. III. Title. IV. Series.

   QH367.5 .S365 1999
   576.8′8–dc21   99-031649

**Visit the Taylor & Francis Web site at
http://www.taylorandfrancis.com**

**and the CRC Press Web site at
http://www.crcpress.com**

Contents

Contributors

Andrew V.Z. Brower is at the Department of Entomology, Oregon State University, Corvallis, OR 97331–2907, USA

Peter L. Forey is at the Department of Palaeontology, The Natural History Museum, Cromwell Road, London SW7 5BD, UK

Julie A. Hawkins is at the Centre for Plant Diversity and Systematics, Department of Botany, School of Plant Sciences, The University of Reading, Reading RG6 6AS, UK

Ian J. Kitching is at the Department of Entomology, The Natural History Museum, Cromwell Road, London SW7 5BD, UK

R. Toby Pennington is at the Royal Botanic Garden Edinburgh, 20a Inverleith Row, Edinburgh EH3 5LR, UK

Paula J. Rudall is at the Royal Botanic Gardens, Kew, Richmond, Surrey TW9 3DS, UK

Robert Scotland is at the Department of Plant Sciences, University of Oxford, South Parks Road, Oxford OX1 3RB, UK

Peter F. Stevens is at the University of Missouri-St Louis, Department of Biology, 8001, Natural Bridge Road, St Louis, MO 63121; and Herbarium, Missouri Botanical Garden, PO Box 299, St Louis, MO 63166, USA

Darrell J. Siebert is at the Department of Zoology, The Natural History Museum, Cromwell Road, London SW7 5BD, UK

Peter H. Weston is at the Royal Botanic Gardens, Sydney, Mrs Macquarie's Road, Sydney, New South Wales 2000, Australia

Ward Wheeler is at the Division of Invertebrate Zoology, American Museum of Natural History, Central Park West @ 79th Street, New York, NY 10024, USA

David M. Williams is at the Department of Botany, The Natural History Museum, Cromwell Road, London SW7 5BD, UK

Preface

This book arises from a symposium of the First Biennial International Conference of the Systematics Association held at Oxford University in August 1997. This meeting, which we hope will be the first of a long series, principally comprised open sessions where systematists were able to present any aspect of their own data and projects. The aim of the Systematics Association Biennial Conferences is to encourage discussion and the exchange of ideas, and particularly to provide a forum for students and young systematists. More specific symposia, such as the one concerning character conceptualization and coding that has led to this book, are embedded within these open sessions. Following the ethos of the general meeting at Oxford, in this symposium we also tried to encourage debate by providing a forum for speakers with contrasting viewpoints, some of which are exploratory, on topics that strike to the heart of systematics.

In addition to the other members of the Conference Organising Committee (Steve Blackmore, Paul Bates and Donald Quicke), we would like to thank the many individuals whose help made the conference possible, particularly Diccon Alexander for producing the stylish artwork of our poster and this book's cover and for several months of administrative work in Edinburgh and Oxford. We also thank Sir Ghillean Prance for giving the conference opening address, Maureen Warwick, Cynthia Styles, Alison Strugnell, Serena Marner and staff of the Oxford University Museum. We thank Jonathan Bennett, Mark Carine, Julie Hawkins and Colin Hughes for invaluable help during the symposium.

Robert Scotland and R. Toby Pennington
Oxford and Edinburgh, 1999

Introduction

R. Toby Pennington

The problem tackled in this book relates to the fact that the outcome of phylogenetic analyses relies upon the constituent parts of a data-matrix. These issues are concerned with character analysis, which Pogue and Mickevich (1990) have referred to as the *bête noire* of systematics. In the first chapter of this volume, Andrew Brower poses the central questions of character analysis: what is a character, and how do we delimit characters, character states, and the relationships between them? Brower emphasizes that this question of character analysis precedes tree-building (by whatever algorithm is used) and thus determines the outcome of a systematic analysis. Despite this importance, much of the literature debating systematic theory over the past three decades has discussed tree-building rather than character analysis. Brower reviews the development of ideas of systematics and homology from pre-Darwinian times to the present, thus placing into context cladistic parsimony (e.g. Farris, 1970, 1983), the tree-building method preferred by the authors in this book.

Most cladists concur with Patterson's (1982) explanation of homology – that it is synapomorphy identified by a cladistic analysis. In simple terms, this means that homology cannot be determined before a cladistic analysis by merely examining structures considered comparable in different organisms; it is the analysis that decides whether comparable structures are homologies or homoplasies. In more phylogenetic terms, the analysis allows us to infer whether structures have evolved once (homology) or more than once (homoplasy). Thus, prior to a cladistic analysis, sytematists should only talk about features that they think might be homologous by constructing hypotheses of 'primary homology' (de Pinna, 1991), which is the core of the problem of character analysis. Brower argues that these hypotheses of primary homology involve several steps that can be viewed as successive stages of character analysis.

The first is the identification of comparable features among the taxa being considered, which allows them to be considered as different manifestations ('states') of the same character. Brower and Schawaroch (1996) described character states as different manifestations of the same character 'topographical identity'. In operational terms this represents the decision as to which characters to include in a data matrix, and how many states each character contains. A cladistic analysis decides whether the same character state shared by different taxa is homologous, thus testing the hypothesis of what Brower and Schawaroch (1996) termed 'character state identity'. However, a cladistic analysis does not test the hypothesis of topographical identity, meaning that it cannot decide whether we have delimited the different characters and character states correctly.

After characters and states have been determined, for each character, decisions have to be made as to how to relate the character states to each other. This operation of defining character state 'order' determines what transformations are permitted between character states. For binary characters (with two states) this is simple as there is only one possibility – one state transforming to the other and vice versa. For characters with more than two states – 'multistate characters' (the most widely used of which is an identified nucleotide position within a DNA/RNA molecule) – the decision is more complex because there are a greater number of possible transformations between states. It is possible to order these state transformations in different ways, and also to weight them differentially (for a more detailed discussion and summary of the literature of ordering and weighting character state transformations, see Kitching *et al.* (1998)). Many workers, especially morphologists, prefer to impose minimal assumptions of how character states can transform (because this amounts to saying one has knowledge of evolution) and thus usually code characters as 'unordered' – i.e. allowing any one state to transform into any other.

Morphology: different ways of coding the same observations

In Chapter 2, Julie Hawkins clearly demonstrates that different systematists perceive and define characters in different ways. She describes 'conventional' as well as nine other different approaches to character coding. The important point is perhaps not the labelling of the different approaches to character coding, but that there are so many ways to code 'different manifestations of the same thing'. The example, taken from Hawkins *et al.* (1997) following Maddison (1993), is for tails, which in different taxa may be red, blue, purple, twisted, straight or absent. This appears simplistic, but is not because these observations show both the complement relation (presence vs absence of tails) and the paired homologue relation (different forms of the tails; see Patterson (1982) for further explanation of this terminology). Hawkins describes how these can be conceptualized as one or more characters, with various numbers of states. She goes on to demonstrate that this is not simply empty theoretical debate by examining character coding in 33 published cladistic data matrices. The clear conclusion is that in the majority of these cases, different coding techniques were employed without discussion. Further examples of different ways to code the same morphological observations are provided by Paula Rudall in Chapter 6.

The logical extension of Hawkins' work is to provide an example where the characters of the same matrix are coded in different ways to discover whether this leads to different phylogenetic conclusions after a cladistic analysis. This is done by Forey and Kitching (Chapter 3), who reanalyse Wake's (1993) data matrix for caecilians (gymnophionians) and a matrix of characters of structures associated with the pupal tongue case of hawkmoths (Sphingidae). In the caecilian example they contrast the results of two cladistic analyses. The first analysis uses 15 multistate morphological characters. The second uses 'presence–absence' coding (also called nominal variable coding (Pimentel and Riggins, 1987) or reductive coding

(Wilkinson, 1995)) which treats every state of each multistate character as a simple presence–absence binary character, resulting in a matrix of 77 binary characters. A simple example contrasting the coding methods is with a single nucleotide sequence character. With multistate coding this has four states, A, G, C, T. With presence–absence coding it becomes four characters: (i) presence or absence of A; (ii) presence or absence of G; (iii) presence or absence of C; (iv) presence or absence of T. The results of the analyses are different. For the caecilian matrix, the multistate data produce eight equally most-parsimonious cladograms, the presence–absence matrix four. Only one of these cladograms was common to the two analyses, and the strict consensus trees shared only two nodes. In the hawkmoth example, five different methods of character coding were applied. Again, the result is clear – different methods of coding the same features result in different phylogenetic conclusions.

Molecules and nucleotide sequence alignment: different ways of coding the same observations

Peter Stevens (Chapter 4) contrasts character analysis in molecular and morphological systematics. As an introduction to this discussion, I will consider what is now the most widely used data in molecular systematics – nucleotide sequence data. For those unfamiliar with molecular systematics, it is worth briefly outlining how these studies proceed. A decision is made as to which small area of the genome (often a single gene) is appropriate for the study. This is based upon knowledge of extent of nucleotide sequence divergence between taxa, based on previous comparative studies. The genomic region is then sequenced for individual organisms chosen to represent the terminal taxa of the analysis. Ideally, these DNA sequences will be the same length (i.e. contain exactly the same number of nucleotides), meaning that each nucleotide position can be numbered, and compared to the same position in a different terminal taxon. However, this is often not the case because the sequences will be different lengths for the different taxa, probably because events of insertion or deletion ('indels') of stretches of DNA have occurred. This is problematic because it is no longer possible to know which nucleotide positions should be compared in different terminal taxa. In this case, sequence alignment is necessary, where 'gaps' (which can be considered as a fifth character state in DNA sequence data, but are more usually scored as missing data, and thus eventually optimized with an A, C, G or T by the parsimony algorithm) are inserted into the DNA sequences to maximize their similarity. These gaps are generally interpreted as representing hypothesized indel events. If sequence alignment is necessary, it determines entirely the delimitation of characters in a molecular systematic study. This important topic, often dealt with too briefly in published papers, is discussed by Ward Wheeler (Chapter 5).

Although molecular and morphological data are analysed by the same algorithms (and indeed can be analysed simultaneously in the same data matrix), the discussion above shows that characters in these data are delimited differently. In a morphological study, characters and their states are delimited first, and then the matrix is constructed by scoring each terminal taxon for the state of each character that it possesses. In contrast, in a nucleotide sequencing study the data are always gathered taxon by taxon as each terminal taxon is sequenced for the genomic region of interest and then sequence alignment determines the exact delimitation of each character. Operationally, if the rows of a matrix are taxa, and each column a character, morphological matrices are often constructed column by column, but nucleotide sequence matrices are always constructed row by row with the exact content of each column determined by sequence alignment (Doyle and Davis, 1998). Viewing things in this simplistic fashion helps clarify the importance of sequence alignment – it is the equivalent of shifting morphological character states sideways for one terminal taxon, which might result in comparing coded states for floral characters for one taxon with those for leaves for another: this would undoubtedly lead to error in subsequent analysis.

Molecules and morphology: overlapping and non-overlapping variation

For a DNA sequence character, there is no debate about the delimitation of the states (A, G, C, T). They are different, discrete and there are no intermediate states. This is a significant advantage over many morphological characters where variation is overlapping. For example, Peter Stevens (Chapter 4) presents 10 characters in the angiosperm genus *Kalmia* (Ericaceae) such as leaf blade (lamina) length and width that show overlapping variation between *Kalmia* species. Such characters are widely used in phylogenetic analyses, despite the uncertainty in the delimitation of character states. In the *Kalmia* study, the measurements of these 10 characters with overlapping variation were graphed in three different ways, and subjects presented with these graphs were asked to delimit character states in the variation of each character. The different ways of graphically presenting the data affected how the character states were delimited by the subjects, emphasizing how complex delimiting character states in such data is likely to be.

Stevens goes on to compare the results of cladistic analyses based upon (i) these morphological characters with overlapping states (using four different matrices reflecting the different state delimitations); (ii) three matrices derived from 23 morphological characters with non-overlapping states; (iii) four matrices combining characters with overlapping and non-overlapping states; and (iv) a matrix of molecular characters. The results of these analyses suggest that states based upon overlapping variation are not as effective in uncovering genealogies as are states based upon non-overlapping variation (see also Stevens, 1991; Gift and Stevens, 1997). Care should thus be exercised when using characters that have overlapping variation in cladistic analyses. When such characters are used, it is better to present the variation explicitly as simple graphs (e.g. Gift and Stevens, 1997), rather than making unsubstantiated assertions based upon visual inspection. Stevens concludes that because single empirical studies in phylogenetics cannot 'prove' anything, the way ahead is more studies where different types of data are produced for each terminal taxon, and then analysed separately and in combination.

Molecules and morphology: reciprocal illumination

Phylogenetic studies using molecular data, particularly nucleotide sequence data, have revolutionized systematics by providing estimates of relationships based upon entirely new data. It has been argued that the most robust hypotheses of relation-ships are derived from simultaneous analysis of all available data – both molecules and morphology (e.g. Bruneau *et al.*, 1995; Nixon and Carpenter, 1996; Pennington, 1996). Some workers do not agree, arguing that if phylogenetic estimates from different datasets are incongruent, at least one of them must be misleading and thus discarded (e.g. Bull *et al.*, 1993). However, what seems recognized less often is that incongruence between datasets might simply result from inadequate char-acter analysis in one of them.

In Chapter 6, Paula Rudall demonstrates how new ideas of relationships based upon molecular data (in this case an analysis of nucleotide sequences of the chloro-plast gene *rbc*L for families of monocotyledons) have resulted in a reassessment of some micromorphological characters. In cases where some micromorphological characters pointed to relationships incongruent with the molecular phylogeny, careful reassessment of these 'cryptic' characters demonstrated that the previous conceptions of primary homology (using the terminology above; note that Rudall uses 'homology' more loosely) were in error. For example, Stevenson and Loconte (1995) coded *Acorus* and *Hydatella* as possessing a perisperm – a seed storage tissue derived from the nucellus – rather than the more usual endosperm. This was one of the synapomorphies linking these genera in a morphological analysis (Stevenson and Loconte, 1995). In contrast, the *rbc*L analysis showed the taxa to be distantly related. This led to a re-examination of the two perisperms that showed them to be structurally and developmentally different (Rudall, 1997; Rudall and Furness, 1997). In *Hydatella* the perisperm is subdermal and multi-layered, whereas in *Acorus* it is entirely derived from the single dermal layer of the nucellus. Thus, micromorphology is not incongruent with the molecular data; the problem was inaccurate assessment of primary homology.

Radical critiques: Rolf Sattler and three-item statement analysis

Peter Weston (Chapter 7) explains that cladistics originated as a radical approach to phylogenetic analysis that challenged existing approaches to systematics and the reconstruction of evolutionary history. Cladistics has now become part of the scien-tific mainstream: for example, 80% of the presentations in the First Biennial Conference of the Systematics Association used or discussed cladistic analysis. The final three chapters of this book make a radical critique of mainstream cladistics.

Weston considers the critique by Rolf Sattler, an influential plant morphologist. This is a critique from outside systematics, because Sattler's research is not concerned with solving systematic problems. Sattler sees plant morphological vari-ation as a continuum: 'In contrast to this categorical view of typical classical plant morphology, continuum morphology acknowledges graduations between typical structures ... From this point of view, homology is a matter of degree ... Intermediates are partially homologous to typical representatives of structural cate-gories' (Sattler, 1996: 577). This differs from the classical view of plant morphology, which reduces the diversity of plant form to mutually exclusive categories such as root, shoot, stem, leaf and trichome. Sattler explains the broad acceptance of the classical model because 'typical' patterns (e.g. the typical simple leaf) occur with great frequency. Thus the great majority of structures can be pigeon-holed.

Sattler's main objection to cladistic analysis is similar to his criticism of classical morphology – delimitation of characters and character states is another exercise in pigeon-holing. He dislikes the view of states of a character showing 1:1 correspondence as transformed versions of each other. This transformation is not direct and is, in principle, unobservable. For example, a blue flower does not directly transform into a red flower because most characters of multicellular organisms are reformed during each generation by developmental processes. Sattler (1994) went on to agree with Hay and Mabberley's (1994) argument that since characters and character states are hypotheses (i.e. ideas) they cannot be said to have participated in phylogenetic transformation, a real process.

It is intriguing that a similar criticism, which also objects to the concept of character state transformation, has arisen from within the discipline of cladistics itself. The protagonists of three item statement (TIS) analysis are developing what they consider a more general model of cladistics, which avoids the need for any transformational hypotheses. TIS analysis, discussed by Robert Scotland in Chapter 8 (and termed there 'three taxon statement (TTS) analysis' and by Dave Williams and Darrell Siebert in Chapter 9, has arisen from the school of 'pattern' or 'transformed' cladistics (Platnick, 1979; Nelson and Platnick, 1981; Patterson, 1982; Brady, 1985) that seeks to remove any circularity of reasoning from systematics by removing any inherent assumptions of evolution. The rationale is that if taxonomic groups are constructed with reference to a particular evolutionary model, they will only reflect the assumptions of that model, and will thus not be explicable by any other pattern-generating process. Protagonists of TIS see conventional parsimony analysis as not having removed an evolutionary model because it requires hypotheses of transformation between the states of a single character. This is regarded as an evolutionary model, albeit a simple one (e.g. for an unordered character, it just says that any state can change into any other). Other cladists would disagree. Weston (Chapter 7) defends this concept of character state transformation within conventional parsimony analysis by describing it as a metaphor – that transformations are figurative relations rather than physical processes.

Another problem highlighted by TIS (see Chapter 8) is that in conventional parsimony analysis, specifying groups requires rooting an unrooted network, which usually needs a prior hypothesis of monophyly of a subset of terminal taxa of the analysis (the ingroup; networks can be rooted using other criteria such as ontogeny, although such data are seldom available; see Weston (1988, 1994)). The TIS approach has its base in the taxic view of homology outlined by Patterson (1982). That is, similarities between taxa (primary homologies using the terminology above) define groups. Taxa that share a primary homology form a group in which taxa that lack this feature do not belong (Scotland, Chapter 8). As such, a hypothesis of homology needs no root, and the TIS approach attempts to abandon the need to define a monophyletic ingroup in order to root a cladistic analysis.

Robert Scotland in Chapter 8 explains how, to some extent, the TIS approach simplifies the decision of how to code complex multistate characters (as discussed in the framework of conventional parsimony analysis by Hawkins in Chapter 2, Forey and Kitching in Chapter 3 and Rudall in Chapter 6). It takes each state of these characters and considers it as a separate grouping character – if the feature (state) is present, this specifies a group. If it is absent, it specifies that the terminal

taxon is not a member of this group (although it does not specify that the terminal taxon lacking the feature is a member of a group defined by lack of the feature, because true absence cannot diagnose a monophyletic group). However, TIS cannot solve all the fundamental problems discussed in the earlier chapters of this book. For example, it is still necessary to align nucleotide sequences before conducting a TIS analysis, and it does not remove the problem of recognizing character states in overlapping variation.

Although TIS does provide a radical new way of using the information from comparative anatomy to construct groups, it is by no means accepted by the majority of working cladists (e.g. Harvey, 1993; Kluge, 1993, 1994; Farris *et al.*, 1995). In Chapter 9, Williams and Siebert reply to these criticisms and deal with technical issues relating to the implementation of TIS.

Weston (Chapter 7) emphasizes that 'cladistics has certainly profited handsomely from interactions with radical movements in the past'. Both TIS and Rolf Sattler's criticisms focus attention upon the inherent assumptions of conventional parsimony analysis. As this type of analysis becomes more widespread, there is a danger that workers will ignore these assumptions, considering they have all been adequately justified in the past. Perhaps the most important function of this book is to remind systematists that they still must consider the assumptions that they are making in every step of a cladistic study, from delimiting the taxonomic characters they will use, to delimiting states in these characters, to coding a data matrix, to tree construction, and then interpreting that tree as a phylogenetic hypothesis.

Acknowledgements

I would like to thank Robert Scotland for inviting me to be involved in organizing the symposium that led to this book and for his comments on this manuscript. I also thank Richard Bateman, Anne Bruneau, Pete Hollingsworth, Matt Lavin and Colin Pendry for their critical comments.

References

Brady, R. (1985) On the independence of systematics, *Cladistics*, 1, 113–126.

Brower, A.V.Z. and Schawaroch, V. (1996) Three steps of homology assessment, *Cladistics*, 12, 265–272.

Bruneau, A., Dickson, E.E. and Knapp, S. (1995) Congruence of chloroplast DNA restriction site characters with morphological and isozyme data in *Solanum* sect. *Lasiocarpa*, *Canadian Journal of Botany*, 73, 1151–1167.

Bull, J.J., Huelsenbeck, J.P., Cunningham, C.W., Swofford, D.L. and Waddell, P.J. (1993) Partitioning and combining data in phylogenetic analysis, *Systematic Biology*, 42, 384–397.

De Pinna, M.C.C. (1991) Concepts and tests of homology in the cladistic paradigm, *Cladistics*, 7, 317–338.

Doyle, J.J. and Davis, J.I. (1998) Homology in molecular phylogenetics: a parsimony perspective, in Soltis, D.E., Soltis P.S. and Doyle, J.J. (eds) *The Molecular Systematics of Plants II: DNA Sequencing*, Dordrecht: Kluwer, pp. 101–131.

Farris, J.S. (1970) Methods for computing Wagner trees, *Systematic Zoology*, 19, 83–92.

Farris, J.S. (1983) The logical basis of phylogenetic analysis, in Platnick, N.I. and Funk, V.A. (eds) *Advances in Cladistics*, vol. 2, New York: Columbia University Press, pp. 7–36.

Farris, J.S., Källersjo, M., Albert, V.A., Allard, M., Anderberg, A., Bowditch, B., Bult, C., Carpenter, J.M., Crow, T.M., De Laet, J., Fitzhugh, K., Frost, D., Goloboff, P., Humphries, C.J., Jondelius, U., Judd, D., Karis, P.O., Lipscomb, D., Luckow, M., Mindell, D., Muona, J., Nixon, K., Presch, W., Seberg, O., Siddall, M.E., Struwe, L., Tehler, A., Wenzel, J., Wheeler, Q. and Wheeler, W. (1995) Explanation, *Cladistics*, **11**, 211–218.

Gift, N. and Stevens, P.F. (1997) Vagaries in the delimitation of character states in quantitative variation – an experimental study, *Systematic Biology*, **46**, 112–125.

Harvey, A.W. (1993) Three-taxon statements: more precisely, an abuse of parsimony? *Cladistics*, **8**, 345–354.

Hawkins, J.A., Hughes, C.E. and Scotland, R.W. (1997) Primary homology assessment, characters and character states, *Cladistics*, **13**, 275–283.

Hay, A. and Mabberley, D.J. (1994) On perception of plant morphology: some implications for phylogeny, in Ingram, D. and Hudson, A. (eds) *Shape and Form in Plants and Fungi*, London: Academic Press, pp. 101–117.

Kitching, I.J., Forey, P.L., Humphries, C.J. and Williams, D.M. (1998) *Cladistics. The Theory and Practice of Parsimony Analysis*, 2nd edn, Oxford: Oxford University Press.

Kluge, A.G. (1993) Three-taxon transformation in phylogenetic inference: ambiguity and distortion as regards explanatory power, *Cladistics*, **9**, 246–259.

Kluge, A.G. (1994) Moving targets and shell games, *Cladistics*, **10**, 403–413.

Maddison, W.P. (1993) Missing data versus missing characters in phylogenetic analysis, *Systematic Biology*, **42**, 576–581.

Nelson, G.J. and Platnick, N.I. (1981) *Systematics and Biogeography: Cladistics and Vicariance*, New York: Columbia University Press.

Nixon, K.C. and Carpenter, J.M. (1996) On simultaneous analysis, *Cladistics*, **12**, 221–241.

Patterson, C. (1982) Morphological characters and homology, in Joysey, K.A. and Friday, A.E. (eds) *Problems in Phylogenetic Reconstruction*, London: Academic Press, pp. 21–74.

Pennington, R.T. (1996) Molecular and morphological data provide resolution at different hierarchical levels in *Andira*, *Systematic Biology*, **45**, 496–515.

Pimentel, R.A. and Riggins, R. (1987) The nature of cladistic data, *Cladistics*, **3**, 201–209.

Platnick, N.I. (1979) Philosophy and the transformation of cladistics, *Systematic Zoology*, **28**, 537–546.

Pleijel, F. (1995) On character coding for phylogeny reconstruction, *Cladistics*, **11**, 309–315.

Pogue, M.G. and Mickevich, M.F. (1990) Character definitions and character state delineation: the *bête noire* of phylogenetic inference, *Cladistics*, **6**, 319–361.

Rudall, P.J. (1997) The nucellus and chalaza in monocotyledons: structure and systematics, *Botanical Review*, **63**, 140–184.

Rudall, P.J. and Furness, C.A. (1997) Systematics of *Acorus*: ovule and anther, *International Journal of Plant Science*, **158**, 640–651.

Sattler, R. (1994) Homology, homeosis and process morphology in plants, in Hall, B.K. (ed.) *Homology, the Hierarchical Basis of Comparative Biology*, New York: Academic Press.

Sattler, R. (1996) Classical morphology and continuum morphology: opposition and continuum, *Annals of Botany*, **78**, 577–581.

Stevens, P.F. (1991) Character states, morphological variation and phylogenetic analysis: a review, *Systematic Botany*, **16**, 553–583.

Stevenson, D.W. and Loconte, H. (1995) Cladistic analysis of monocot families, in Rudall, P.J., Cribb, P.J., Cutler, D.F. and Humphries, C.J. (eds) *Monocotyledons: Systematics and Evolution*, Kew: Royal Botanic Gardens, pp. 543–578.

Wake, M.H. (1993) Non-traditional characters in the assessment of caecilian phylogenetic relationships, *Herpetological Monographs*, **7**, 42–55.

Weston, P.H. (1988) Indirect and direct methods in systematics, in Humphries, C.J. (ed.) *Ontogeny and Systematics*, New York: Columbia University Press, pp. 27–56.

Weston, P.H. (1994) Methods for rooting cladistic trees, in Scotland, R.W., Siebert, D.J. and Williams, D.M. (eds) *Models in Phylogeny Reconstruction*, Oxford: Clarendon Press, pp. 125–155.

Wilkinson, M. (1995) A comparison of two methods of character construction, *Cladistics*, **11**, 297–308.

Chapter 1

Homology and the inference of systematic relationships: some historical and philosophical perspectives

Andrew V. Z. Brower

Introduction: the evolution of systematics

> ... systematists always have been, are, will be, and should be, typologists.
> (Gareth Nelson and Norman Platnick, 1981: 328)

In 1995, Sneath published an account of the struggle between alternative schools of systematic inference, arguing that Hennigian phylogenetics was effectively dead and that 'numerical taxonomy' had finally triumphed in the war of opposing methodologies. More recently, De Queiroz and Good (1997) have published a historical review of phenetic clustering methods, in which they arrive at the opposite conclusion, announcing the victory of phylogenetics and ascribing the demise of numerical taxonomy to its incompatibility with a view of systematics based on common ancestry. It is striking that scholars whose profession is to infer historical patterns from fragmentary evidence can arrive at such diametrically opposed interpretations of relatively recent and well-documented events. I do not intend to dissect the incompatible views of history presented by these authors in this essay. Instead, the purpose of this introduction is to offer a third alternative account of the development of current opinions on systematic inference.

A major goal of this chapter is to develop the argument that systematists should view the construction of natural classifications as an empirical, nominalist operation. I will criticize as superfluous to the success of the systematic endeavour both the reliance on assumptions that evolution has occurred and the popular belief that clades are explained by the process of phylogeny. The relationship between pattern and process in systematics will be clarified by investigating the methods of doing systematics with an eye to the historical development of ideas and technologies.

To begin, we need to look back to the 18th century, to pre-Darwinian, pre-evolutionary times. Since the Renaissance, ships had been sailing around the world, bringing back strange new plants and animals to European taxonomic encyclopaedists of the Enlightenment such as Gesner, Ray and Linnaeus. Cataloguing and naming these wondrous creations was a major industry, and keeping them organized became increasingly complicated. As collections grew, it became obvious that these strange new specimens were not all completely strange, and not utterly new. Many of them shared features with familiar plants and animals from Europe that suggested a common plan or relationship, and it became increasingly clear that a recognizable hierarchy of groups could be formed. Further, fossil skeletons of unknown creatures could be rearticulated, based on their similarity to the structures of living animals. For example, when the amateur palaeontologist Reverend W.D. Conybeare (1825) announced the discovery (by Mary Anning) of an almost fully articulated fossil plesiosaur, he could boast that

> At the period of my former communications it was natural and even just that in the minds of many persons interested in such researches, much hesitation should be felt in admitting the conclusion of an observer who was avowedly inexperienced in comparative anatomy; and there might have then appeared reasonable ground for the suspicion that . . . I had been led to constitute a fictitious animal from the juxtaposition of incongruous members, referable in truth to different species. But the magnificent specimen recently discovered at Lyme has confirmed the justice of my former conclusions in every essential point connected with the organization of the skeleton.

This anecdote illustrates the predictive power of comparative morphology and the effectiveness of homology assessment in the absence of an evolutionary worldview. More than 30 years prior to the publication of Darwin's *Origin of Species* (1859), a minister who had no notions about common ancestors was able to take a heap of disarticulated fossil bones from an animal never before seen, and reconstruct a skeleton that matched almost perfectly with an intact skeleton discovered subsequently in the same fossil bed. (It is worth digressing to point out that the reason we recognize that any fossil is the remains of a once-living organism at all is not because anyone ever saw an extant dinosaur or sabre-toothed cat, or ammonoid, but because of the structural similarities between the fossils and the skeletons of organisms that are alive today.)

The explanation given by pre-Darwinian naturalists for the similarities among taxa was that the underlying pattern reflected God's plan of creation. Would it be possible to figure out that plan? This was the great challenge for 'philosophical anatomists' of the late 18th and early 19th centuries, such as Goethe and von Baer in Germany, Bonnet, Lamarck, Geoffroy St Hilaire and de Blainville in France, and Macleay, Westwood, Strickland and Owen in Britain.

It is no coincidence that the same criteria (similarity and correspondence of relative position) that we use today for homology recognition were described by Etienne Geoffroy St Hilaire in 1818. The great 1830 debate between Geoffroy and Cuvier before the Academy in Paris was not over the utility of 'analogies', but whether all animals had been constructed on a single fundamental plan that could be inferred by comparison of their structures. At the time, Cuvier's sceptical empiricism (and political might) diminished Geoffroy's idealistic philosophical scheme in the opinions of the scientific community in France and abroad. However, subsequent 'positive facts', as Cuvier would have called them, or corroborated empirical observations, as we would call them now, have borne out Geoffroy's position: the empirical evidence implies a single Natural System.

Modern concepts of homology are all manifestations of our conviction that a single Natural System is the appropriate organizational scheme for thinking about biological diversity. The claim that homology is defined as similarity due to common descent is currently popular because the Natural System can be explained by the theory of evolution. But the empirical observation of similar features among taxa, and the further observation that these features imply a pattern of grouping into nested sets based on the similarities and differences of these features, are not dependent on evolution, God's plan, or any other explanatory theory. These observations constitute the explanandum — the independent, empirical thing to be explained. In fact, as Brady (1985: 117) has argued, the independence of the pattern from explanatory process theories is a necessary feature of a scientific approach to systematics:

> By making our explanation into the definition of the condition to be explained, we express not scientific hypothesis but belief. We are so convinced that our explanation is true that we no longer see any need to distinguish it from the situation we are trying to explain. Dogmatic endeavours of this kind must eventually leave the realm of science.

A convenient point at which to resume the historical narrative is the founding of the Systematics Association 60 years ago, at the time of the Modern Synthesis (see Mayr and Provine, 1980; Smocovitis, 1992). Those were heady times for evolutionary biology, but the rise of evolutionist hegemony began a dark age for systematics that continued for at least 20 years, and is arguably ongoing (Wheeler, 1995). Once regarded as 'the queen of sciences', systematics came to be viewed as dry, intellectually sterile pigeon-holing of specimens, with little theoretical insight to offer to the modern age. In 1940, Julian Huxley heralded a 'new systematics' which redefined the fundamental problem of systematics from discovering the hierarchy of nature to 'detecting evolution at work' (1940: 2). Huxley, as he admitted himself (1940: v), was not a systematist.

By 1960, none of the leading authorities on systematics were systematists. Instead, they were population biologists such as Ernst Mayr and palaeontologists like G. G. Simpson, with outlooks and scientific agendas that abandoned the traditional systematic goals of identifying, naming and inferring phylogenetic relationships among taxa. Thus, Simpson (1961: 7) could define systematics as 'the scientific study of the kinds and diversity of organisms and of any and all relationships among them', a definition so broad as to be virtually meaningless (it would seem to encompass both ethology and ecology), while Mayr (1963) could ignore it entirely in his *magnum opus*, *Animal Species and Evolution*.

With the rise of phenetics in the 1960s and of cladistics in the 1970s, interest was renewed in the philosophical and methodological underpinnings of systematics in the original, narrow sense (Sokal and Sneath, 1963; Hennig, 1966; Blackwelder, 1967). Inklings of the phenetic worldview can be observed in the empiricist writings of Woodger (1937), Gilmour (1940) and Gregg (1950), but it was not until the advent of computers that the numerical taxonomic paradigm began to blossom. The mechanical ability to perform the complex calculations necessary to infer hierarchical patterns from character data meant that systematic controversies once resolved by authority could be tackled with objective and repeatable algorithms. With these technological advances, controversy immediately erupted over how systematic evidence should be interpreted: once some data have been gathered, how should they be used to infer phylogenetic relationships?

Three basic approaches emerged that can be 'keyed out' by Aristotelian logical division as follows: first, should systematists rely upon the traditional intuitive approach or upon some mechanical process to build phylogenetic trees? These alternatives represent the distinction between traditional systematics (e.g. Mayr, 1969) and what Sokal and Sneath (1963) called numerical taxonomy. Numerical taxonomy is an intellectual child of J.S. Mill (1843) and the philosophy of logical positivism that was popular in the first half of the 20th century. The goal of the logical positivists was to banish metaphysical statements (statements about the true condition of things irrespective of human cognition) from the explanation of phenomena. Numerical taxonomists, disgusted with the metaphysical excesses and authoritarian style of evolutionary systematists, wanted the same thing: a systematics based on empirical observation and intersubjectively operational methods that could be repeated by 'an intelligent ignoramus' (Sokal and Sneath, 1966). Their data were derived from observation of the phenotype and their taxa were clustered by repeatable algorithms. The entire process was conducted without reference to phylogenetic speculation, hence the common dialectic of phenetics vs phyletics (later 'cladistics', thanks to Mayr's popularization of Cain and Harrison's (1960) term).

If we believe that an algorithmic approach to systematic inference is appropriate, a second choice is to pick a criterion for grouping: by which of the infinite number of possible algorithms should the data be interpreted? The phenetic–cladistic debate of the 1970s focused on whether the computer-generated trees should be based on a metric of overall similarity of character states among taxa, or on a quantification of the character state changes among taxa. To visualize this distinction, imagine a set of taxa that display two character states (A and B) for some character. A phenetic interpretation of these data is that the two character states imply two mutually exclusive groups, Group A and Group B. The cladistic interpretation of the same information is that the data implies only one exclusive group, either Group A or Group B, nested within the other. This is because the change of the character from one state to the other, and not the existence of distinct states, constitutes the evidence for grouping. A parsimonious reconstruction of state transformations demands one state change from A to B or B to A. Thus, one group of taxa is united by exhibiting the changed state, while the taxa exhibiting the other state are not united with one another by any evidence that does not also unite them with the taxa exhibiting the alternate, derived state. This difference represents the fundamental operational distinction between phenetic and cladistic methodologies.

From an operational perspective, the great advance for cladistics was Farris's implementation of Hennig's rather metaphysically based phylogenetic ideas in an explicit numerical procedure (Farris, 1970; Farris *et al.*, 1970) that made cladistics competitive with computerized phenetic methodologies. While evolutionary systematists (and Hennig) had no method to realize their phylogenetic theories, and pheneticists had no theory to discriminate among the many possible methods of grouping based on similarity, cladists, particularly pattern cladists who saw fit to separate and discard the metaphysical Hennigian husk of common ancestry, were able to combine the advantages of both, using the method of grouping by parsimonious patterns of shared character state change.

The trend during the 1960s and 1970s was thus towards more and more direct treatment of character information via a transparent and philosophically justifiable analytical procedure. More recently, with the rise of DNA sequencing and the invasion of systematics by population geneticists and molecular biologists ignorant of systematic theory, the controversy has shifted to ask if character information should be corrected by differential weighting, using models based on generalized rates of molecular evolution or other ideas about the informativeness of some characters relative to others. It is ironic that the 'correction' of the data by differential weighting, the central issue of Mayr's (1969) authoritarian dispute with cladistics and phenetics, is again at the core of the disagreement between modern cladists and advocates of maximum likelihood methods.

Character analysis and phylogenetic inference – separate but equal

> Homology is an inference. It is the inference that a given similarity is the outcome of common inheritance rather than of other influences making for similarity. As an inference, it is based on the weighting and interpretation of evidence. We cannot avoid the difficulties of making a judgement in a particular case, all we can hope to do is to clarify the basis on which that judgement is to be made.
>
> (Etkin and Livingston, 1947: 472)

The topic of this book is a question to which Pogue and Mickevich (1990) have referred as the *bête noire* of systematics: what is a character, and how do we delimit characters, character states, and the relationships between them? Character analysis has long been recognized as a distinct procedure that is prior to tree construction by morphologists, and alternative methods of scoring the same features in the data matrix can have dramatic effects on the resultant topology, even if the same tree-building method is used (see Pimentel and Riggins, 1987; Pleijel, 1995; and Chapters 2 and 3). A dramatic morphological example of the effect of insufficient attention to data matrix construction is described by Patterson and Johnson (1997).

At the core of the character analysis problem is the problem of what de Pinna (1991) has called primary homology, which is the theory underlying the empirical basis of systematics. Most readers will be familiar with Patterson's (1982) equation of homology with synapomorphy. As de Pinna noted in his review, synapomorphies are end-products of cladistic analysis, distinguished from convergence by cladogram construction and from symplesiomorphies by rooting. Prior to constructing the tree, therefore, only similarities and differences in features exist among the taxa being compared, and homology in the Pattersonian sense remains uncertain. Systematists therefore need an alternative terminology to talk about features that we think might be homologous, but have not yet been tested by congruence with the other characters in a cladistic analysis. We could just call it 'similarity', as set out in Remane's (1952) criteria (which basically reiterate Geoffroy–St Hilaire's Principle of Connections and Principle of Composition). In de Pinna's terminology, this is referred to as primary homology, but Val Schawaroch and I (1996) have argued that a statement of primary homology involves several different things, which can be viewed as successive stages of character analysis.

First, one needs to find comparable features among the taxa being considered. This largely intuitive process has been accomplished in the same way for hundreds of years. If one thinks of the steps involved in building a data-matrix, this is the stage in which the observer delineates the characters that she intends to score (i.e. enumerates a row of different characters across the top of the data matrix). Each of these characters is considered to be independent, for the sake of the subsequent analysis. We called the similarity of features among organisms that allows them to be considered as different manifestations of the same character 'topographical identity', following the terminology from Jardine (1969). Topographical identity corresponds to Patterson's (1982) transformational homology, because no hierarchical patterns of grouping are implied by the identification of topographically identical features. A subsequent stage to identification of topographically identical features is the parsing of variation of the features into discrete states (which corresponds to scoring values in the individual cells of the data matrix). Features among taxa will be identified as the same or different states. We called this sameness 'character state identity'. Character state identity, but not topographical identity, is tested by character congruence in cladistic analysis. After characters and states have been determined, character states may be ordered, or linked together into transformation series that differentially weight the various possible state transformations. Just as it is not possible to know true patterns of phylogeny, however, so it is not possible to know true transformation series among character states, or to know with certainty where one character ends and the next begins. Because of these limitations to knowledge, many morphologists prefer to employ transformation schemes that impose minimal *ad hoc* assumptions about differential weights (e.g. unordered or presence–absence coding).

As pointed out in Brower and Schawaroch (1996), DNA characters are treated the same as morphological characters in this stepwise scheme. Topographical identity of nucleotide sites is provided by alignment of the sequences, which may be extremely ambiguous (Gatesy *et al.*, 1993; see Chapters 4 and 5). Character state identity is trivial once an alignment is produced: A, G, C, T and gap are distinct and unambiguous states for a given nucleotide site. Again, the individual character state identities are tested by cladistic analysis, but not the alignment, which is an *a priori* statement of topographical identity among the sequences being compared. Once sequences are aligned, they are tested by character congruence in cladistic analysis in exactly the same way as morphological characters.

Together, the steps outlined above yield the hypotheses of putative synapomorphy that de Pinna (1991) called primary homology. The point here is that as the building blocks of systematic hypotheses, characters are distinct and separate things requiring methodical consideration prior to phylogenetic analysis (tree building). I am not suggesting that reciprocal illumination is not possible – the stepwise procedure outlined here may be corrected as new information bears on particular character interpretations – but it is imperative to appreciate the logical priority of character analysis to phylogenetic analysis. Unless a clear understanding of characters as observations with a unitary contribution to the data matrix is shared among systematists, the edifice of systematics becomes a Tower of Babel, in which no two people speak the same systematic language and no two classifications are comparable. Such misunderstanding of characters, homology, and their respective roles in phylogenetic analysis opens the door for a wide variety of ill-conceived methods of topological inference that further muddy the waters of the systematic endeavour.

Although the merits of alternative tree-building methods are a topic for another symposium, I will digress momentarily to offer one prognostication: the maximum likelihood approach to phylogenetic inference will founder in multiplicity in the same way that phenetics did, not because it lacks an optimality criterion, but because its optimality criterion is conditional upon very detailed prior deterministic knowledge. It is very doubtful that agreement can be reached on an appropriate model or meta-model of evolutionary change to front-load the likelihood algorithm. Even if such a model were agreed upon, likelihood would still be only a much more complex and not demonstrably superior alternative to cladistic parsimony. It is possible to evaluate empirically the relative accuracy of different phylogenetic methods, but such experiments are not meaningful because they require prior knowledge of the phylogeny being tested. As Sober (1993) has pointed out, such systematic modelling is a waste of time, because extrapolation from a test case provides very little prediction of the accuracy of a method beyond the trivial realm of the test itself. On the other hand, if total knowledge of the validity of an evolutionary model were attainable, it would vitiate the purpose of systematic analysis in the first place (Platnick, 1979). If we believe that there is one hierarchical pattern that reflects the diversity of life on this planet, a common language is fundamental to the endeavour of describing its shape, and that means that there should be a single standard procedure for analysing data and inferring phylogenetic relationships. This is a basic reason why it is illogical to try multiple methods, in the hope that they won't disagree. Multiple tree-building methods no more corroborate the truth of a given topology than multiple statistical tests corroborate the results of an experimental study.

Nominalism vs essentialism

> Homologous features (or states of features) in two or more organisms are those that can be traced back to the same feature (or state) in the common ancestor of those organisms.
>
> (Ernst Mayr, 1969: 85)

> Homology has often been defined in an 'evolutionary' sense, and 'homologous' features have been said to be 'traceable' back to the 'same' (i.e., 'homologous') feature of some common ancestor. Operationally, common ancestors are at best only hypothetical constructs. Thus, 'tracing homologous features back to some common ancestor' amounts only to erecting an hypothesis of ancestral conditions.
>
> (Gareth Nelson, 1970: 378)

> Taxa are monophyletic if and only if they share a common ancestor, irrespective of evidence or belief.
>
> (Michael Ghiselin, 1984: 109)

A basic tension in science is the problem already mentioned, that as observers, scientists are severely limited in their access to truth. We can make observations, but for science to progress, general propositions must be advanced by a process of inductive inference, which has been viewed as philosophically problematical for some time (Hume, 1779; Popper, 1965). Thus, all scientists to a greater or lesser extent conduct their enquiries in an atmosphere of suspended disbelief. Whether or not they choose to admit that they are doing so is an important distinction, the distinction I consider to represent the difference between nominalism and essentialism. (One could substitute empiricism for nominalism, and realism for essentialism, but the contrast I want to draw would be the same – definitions are arbitrary and need not be orthodox, as long as they are clear.)

In my view, these 'isms' represent alternative ways of thinking about the world. The essentialist point of view focuses on ontology – things as they actually are ('irrespective of evidence or belief') and maintains that truth is an accessible goal of the scientific endeavour. Having suspended disbelief in the limitations of inductive inference, the essentialist sees science as revealing facts and building general theories or 'laws' about the real world that may subsequently be used as premises for deductive reasoning. Nominalism is the opposite, focusing on the structure of knowledge, and explicitly recognizing that observations are intrinsically mediated by theoretical constraints that limit our access to 'reality'. The goal of nominalist science is to erect hypotheses that are simple and consistent with the available data. Simplicity is fundamental, because, in the absence of verifiability, it provides an epistemological lower limit to the range of possible explanations of phenomena – phenomena are not necessarily simple, but a simple and sufficient explanation of a phenomenon is better than a more complicated one. Scientific knowledge grows in the nominalist view because observations that are repeated often enough become uncontroversial, and are therefore accepted as background knowledge for subsequent enquiry. Science cannot advance without assuming some background knowledge, but just how much should be taken for granted is problematical. Because of its reliance on simplicity as an epistemological axiom, the nominalist viewpoint is that background knowledge should be assumed parsimoniously – less is better.

There are many possible ways to do systematics, given this nominalist perspective. Why should we choose to group things the way we do, instead of some other way? In the absence of truth, the best method for grouping things is that which results in groups with the highest information content and the greatest ability to explain additional observations. To nominalists such as Mill (1843), Gilmour (1940) and Popper (1965), a natural classification is one which offers the highest explanatory content about the things being grouped. Farris (1979; 1983) has argued compellingly that cladistic parsimony is the method that yields groups that are the most natural in this sense, and has explicitly tied these arguments to the notion of simplicity (Farris, 1982).

What provides the empirical raw material for taxonomic grouping in a nominalist worldview? If observations are theory-laden, and thus are not pure 'facts', why should a nominalist believe anybody else's data, or even his own eyes? As long as different observers can agree on what has been observed, the theory-laden basis of observation is not a serious problem: intersubjectivity – the agreement among observers as to what has been observed – is a useful surrogate for objectivity. As Popper (1965: 111) stated,

> The empirical basis of objective science has ... nothing 'absolute' about it. Science does not rest upon rock bottom. The bold structure of its theories rises, as it were, above a swamp. It is like a building erected on piles. The piles are driven down from above into the swamp, but not down to any natural or

'given' base; and when we cease our attempts to drive our piles into a deeper layer, it is not because we have reached firm ground. We simply stop when we are satisfied that they are firm enough to carry the structure, at least for the time being.

Everybody sees himself as a nominalist, and accuses somebody else of being an essentialist. Ernst Mayr has written a lot about the concept of essentialism in evolutionary biology (e.g. Mayr, 1982), using it as a disparaging epithet to describe the systematic philosophy of traditional taxonomists who recognized taxa on the basis of characters. Mayr argued that the concept of character is a vestige of Platonic typological thought – that characters are features that reflect the ideal, eternal essences of species. Dismissing this idea, Mayr claimed that essentialist typological thinking has been replaced by modern 'population thinking', which emphasizes variation at all levels and commonality at none. In my view (see also Wheeler, 1995), Mayr's characterization of this distinction represents perhaps the most profound misunderstanding of the systematic endeavour that has ever been promulgated, because he has merely substituted his own non-operational essentialistic views about biological species for what he perceives to be those of traditional systematists.

Mayr's argument is objectionable on both practical and theoretical grounds. From a practical perspective, the fundamental problem of systematics is to discover orderly pattern in nature, which we believe to be usefully represented as an irregularly branching hierarchy (this assumption is rarely formally tested, but its repeated empirical corroboration has led Platnick (1982) and Brady (1985) to argue that the existence of a hierarchy is an empirical fact). Fact or not, it is difficult to imagine how the inference of hierarchical pattern could be accomplished without the comparison of similarities and differences of features of organisms. Inevitably, the observer is forced to select physical characteristics of representative organisms, or *characters*, that allow the taxa that the organisms represent to be sorted into groups within groups. No matter what criterion is used to recognize group membership, the procedure of empirical grouping demands observation of characters that allow discrimination of members of one group from another. There is no implication that these features necessarily represent anything 'real' or essential about the things being grouped, nor that the groups themselves are real in the Platonic sense. Empirical philosophers since Locke have recognized the inaccessibility of the true nature of things from observation. Holomorphology, being the sum total of observed features of taxa (including DNA), provides the ontological mud of our systematic swamp. The reason we describe our methods clearly and publish our data matrices is precisely so that others may intersubjectively corroborate those methods and data with repeated or additional observations.

Truth, except as a figure of speech, does not exist in empirical science. The phylogenetic groups we recognize are merely hypotheses that are useful because they provide a maximally informative summary of the empirical pattern that is supported by observations at an intersubjectively uncontroversial level of organization. Because the metaphysical goal of reconstructing the 'true' phylogeny has been frankly and explicitly abandoned in favour of development of and adherence to the most logically sound methodology, the procedure of systematic inference practised by cladists is not essentialist, but nominalist (Ghiselin's mistaken views notwithstanding). Under this approach, it becomes obvious that notions of phylogenetic truth and accuracy are will-o'-the-wisps, because tests of such ideas are inevitably dependent on the plausibility of the original observations by which the criterion of correctness is set. The corroboration of hypotheses is the closest to reality that empirical science ever comes.

To sum up, I have argued that systematics is a basically nominalist operation, and that accusations of essentialism practised by pre-Darwinian or modern empirical systematists are misplaced because they merely criticize one ontological viewpoint from the perspective of another, when all such explanations are equally irrelevant to the discovery of the pattern. The basic process of recognizing homologous features has not changed for 200 years or more, and although we now have more precise analytical tools and a wider range of characters to sample, there is nothing fundamentally different about the way we infer the pattern of the Natural System from the way it was inferred by our predecessors. I re-emphasized points made by Brady (1985) and Rieppel (1988) that assumptions of background knowledge should be minimized to make systematic hypotheses boldest, and this includes the unnecessary assumption of evolutionary theories and models. Because all empirical research is based on observation, science necessarily rests on some foundation of background knowledge which is in part theoretical and essential (topographical identity is the theoretical claim of essential similarity between different physical entities). But in order for scientific hypotheses to be bold, they should rely as little as possible on extraneous supplementary theories that dilute the testability of the underlying data. This is just the old principle of parsimony: events in the real world may not occur parsimoniously, but parsimony is the epistemological limit that should be striven for in their interpretation.

Acknowledgements

I thank Toby Pennington, and particularly Robert Scotland for efforts to organize the symposium and its book, generous hospitality, and indulgence of my procrastination and pontification. Thanks also to Darlene Judd and Olivier Rieppel for discussions or comments on the manuscript. I was supported in part by the Smithsonian Institution and the Systematics Association. Finally, I acknowledge my debt to Ron Brady, Steve Farris, Gary Nelson, Colin Patterson, Norman Platnick and the other cladists, and to the pre-Darwinians, for their ideas.

References

Blackwelder, R.E. (1967) *Taxonomy*, New York: Wiley.
Brady, R.H. (1985) On the independence of systematics, *Cladistics*, 1, 113–126.
Brower, A.V.Z. and Schawaroch, V. (1996) Three steps of homology assessment, *Cladistics*, 12, 265–272.
Cain, A.J. and Harrison, G.A. (1960) Phyletic weighting, *Proceedings of the Zoological Society of London*, 135, 1–31.
Conybeare, W.D. (1825) On the discovery of an almost perfect skeleton of the *Plesiosaurus, Philosophical Magazine*, 65, 412–421.
Darwin, C. (1859) *On the Origin of Species*, London: John Murray.

De Pinna, M.C.C. (1991) Concepts and tests of homology in the cladistic paradigm, *Cladistics*, 7, 367–394.

De Queiroz, K. and Good, D.A. (1997) Phenetic clustering in biology: a critique, *Quarterly Review of Biology*, 72, 3–30.

Etkin, W. and Livingston, L.G. (1947) A probability interpretation of the concept of homology, *The American Naturalist*, 81, 468–473.

Farris, J.S. (1970) Methods for computing Wagner trees, *Systematic Zoology*, 19, 83–92.

Farris, J.S. (1979) On the naturalness of phylogenetic classification, *Systematic Zoology*, 28, 200–214.

Farris, J.S. (1982) Simplicity and informativeness in systematics and phylogeny, *Systematic Zoology*, 31, 413–444.

Farris, J.S. (1983) The logical basis of phylogenetic analysis, in Platnick, N.I. and Funk, V.A. (eds) *Advances in Cladistics*, vol. 2. New York: Columbia University Press, pp. 7–36.

Farris, J.S., Kluge, A.G. and Eckhardt, M.J. (1970) A numerical approach to phylogenetic systematics, *Systematic Zoology*, 19, 172–191.

Gatesy, J., DeSalle, R. and Wheeler, W. (1993) Alignment-ambiguous nucleotide sites and the exclusion of systematic data, *Molecular Phylogenetics and Evolution*, 2, 152–157.

Ghiselin, M.T. (1984) "Definition," "character," and other equivocal terms, *Systematic Zoology*, 33, 104–110.

Gilmour, J.S.L. (1940) Taxonomy and philosophy, in Huxley, J.S. (ed.) *The New Systematics*, Oxford: Oxford University Press, pp. 461–474.

Gregg, J.R. (1950) Taxonomy, language and reality, *The American Naturalist*, 84, 419–435.

Hennig, W. (1966) *Phylogenetic Systematics*, Urbana, IL: University of Illinois Press.

Hume, D. (1779) *Dialogues Concerning Natural Religion*, Hafner Library of Classics edn, 1948, New York: Hafner Press.

Huxley, J.S. (1940) Towards the New Systematics, in Huxley, J.S. (ed.) *The New Systematics*, Oxford: Oxford University Press, pp. 1–46.

Jardine, N. (1969) The observational and theoretical components of homology: a study based on the morphology of the dermal skull-roofs of rhipidistian fishes, *Biological Journal of the Linnean Society*, 1, 327–361.

Mayr, E. (1963) *Animal Species and Evolution*, Cambridge, MA: Belknap Press.

Mayr, E. (1969) *Principles of Systematic Zoology*, New York: McGraw-Hill.

Mayr, E. (1982) *The Growth of Biological Thought*, Cambridge, MA: Belknap Press.

Mayr, E. and Provine, W.B. (eds) (1980) *The Evolutionary Synthesis*, Cambridge MA: Harvard University Press.

Mill, J.S. (1843) Of classification, as subsidiary to induction, in *A System of Logic – Ratiocinative and Inductive*, London: Longman, pp. 465–479.

Nelson, G.J. (1970) Outline of a theory of comparative biology, *Systematic Zoology*, 19, 373–384.

Nelson, G. and Platnick, N. (1981) *Systematics and Biogeography*, New York: Columbia University Press.

Patterson, C. (1982) Morphological characters and homology, in Joysey, K.A. and Friday, A.E. (eds) *Problems of Phylogenetic Reconstruction*, London and New York: Academic Press, pp. 21–74.

Patterson, C. and Johnson, G.D. (1997) The data, the matrix, and the message: comments on Begle's 'Relationships of the osmeroid fishes', *Systematic Biology*, 46, 358–365.

Pimentel, R.A. and Riggins, R. (1987) The nature of cladistic data, *Cladistics*, 3, 201–209.

Platnick, N. I. (1979) Philosophy and the transformation of cladistics, *Systematic Zoology*, 28, 537–546.

Platnick, N.I. (1982) Defining characters and evolutionary groups, *Systematic Zoology*, 31, 282–284.

Pleijel, F. (1995) On character coding for phylogeny reconstruction, *Cladistics*, **11**, 309–315.

Pogue, M.G. and Mickevich, M.F. (1990) Character definitions and character state delineation: the *bête noire* of phylogenetic inference, *Cladistics*, **6**, 319–361.

Popper, K.R. (1965) *The Logic of Scientific Discovery*, New York: Harper Torchbooks.

Remane, A. (1952) *Die Grundlagen des natürlichen Systems, der vergleichenden Anatomie und der Phylogenetik*, Leipzig: Acad. Verlagsges.

Rieppel, O.C. (1988) *Fundamentals of Comparative Biology*, Basel: Birkhäusen.

Simpson, G.G. (1961) *Principles of Animal Taxonomy*, New York: Columbia University Press.

Smocovitis, V.B. (1992) Unifying biology: the Evolutionary Synthesis and evolutionary biology, *Journal of the History of Biology*, **25**, 1–65.

Sneath, P.H.A. (1995) Thirty years of numerical taxonomy, *Systematic Biology*, **44**, 281–298.

Sober, E. (1993) Experimental tests of phylogenetic inference methods, *Systematic Biology*, **42**, 85–89.

Sokal, R.R. and Sneath, P.H.A. (1963) *Principles of Numerical Taxonomy*, San Francisco: W. H. Freeman and Co.

Sokal, R.R. and Sneath, P.H.A. (1966) Efficiency in taxonomy, *Taxon*, **15**, 1–21.

Wheeler, Q.D. (1995) The 'Old Systematics': classification and phylogeny, in Pakaluk, J. and Slipinski, S.A. (eds) *Biology, Phylogeny and Classification of Coleoptera: Papers Celebrating the 80th Birthday of Roy A. Crowson*, Warsaw: Muzeum i Instytut Zoologii PAN, pp. 31–62.

Woodger, J.H. (1937) *The Axiomatic Method in Biology*, Cambridge: Cambridge University Press.

Chapter 2

A survey of primary homology assessment: different botanists perceive and define characters in different ways

Julie A. Hawkins

Introduction

Cladistic analyses begin with an assessment of variation for a group of organisms and the subsequent representation of that variation as a data matrix. The step of converting observed organismal variation into a data matrix has been considered subjective, contentious, under-investigated, imprecise, unquantifiable, intuitive, as a black-box, and at the same time as ultimately the most influential phase of any cladistic analysis (Pimentel and Riggins, 1987; Bryant, 1989; Pogue and Mickevich, 1990; de Pinna, 1991; Stevens, 1991; Bateman *et al.*, 1992; Smith, 1994; Pleijel, 1995; Wilkinson, 1995; Patterson and Johnson, 1997). Despite the concerns of these authors, primary homology assessment is often perceived as reproducible. In a recent paper, Hawkins *et al.* (1997) reiterated two points made by a number of these authors: that different interpretations of characters and coding are possible and that different workers will perceive and define characters in different ways. One reviewer challenged us: did we really think that two people working on the same group would come up with different data sets? The conflicting views regarding the reproducibility of the cladistic character matrix provoke a number of questions. Do the majority of workers consistently follow the same guidelines? Has the theoretical framework informing primary homology assessment been adequately explored? The objective of this study is to classify approaches to primary homology assessment, and to quantify the extent to which different approaches are found in the literature by examining variation in the way characters are defined and coded in a data matrix.

Approaches to primary homology assessment

What is a character?

Primary homology assessment is a two-step process, comprising topographic identity and character state identity determination steps (Brower and Schawaroch, 1996). Operationally, these two steps of data matrix construction comprise the recognition of characters, which define columns in the data matrix, and the partitioning of characters into the entries within the column. Although there is a methodological assumption in cladistic analysis that character states are discrete, it is often the case, particularly when analysing species relationships, that many of the features that can be used to separate taxa show continuous variation (Felsenstein, 1988; Chappill, 1989). Given the paucity of qualitative characters and abundance of quantitative variation, many workers have discussed the use of continuous variation in cladistic analyses (Archie, 1985; Pimentel and Riggins, 1987; Felsenstein, 1988; Chappill, 1989; Farris, 1990; Stevens, 1991; Thiele, 1993; Strait *et al.*, 1996; see Chapter 4). The extent to which the literature shows inconsistencies in the coding of quantitative variation as character states has already been examined in detail (e.g. Stevens, 1991). Quantitative variation and character state identity decisions are not pursued further here (see Chapter 4). Rather, the approaches to primary homology assessment which define the relationship between characters and character states in different ways are reviewed. These are different approaches to the question 'what is the character?'.

Platnick (1979: 542) presented a clear definition of the term character: 'a character consists of two or more different attributes (character states) found in two or more specimens that, despite their differences can be considered alternate forms of the same thing (the character)'. De Pinna (1991) clarified this issue, noting that character states are attributes that can be proposed as transformations of each other, while characters are putatively independent from one another. This concept of the character provides the basis for what is referred to here as conventional character coding, following Hawkins *et al.* (1997). Using coding examples for an attribute (tails – following Maddison, 1993) which may be red, blue, purple, twisted, straight or absent, some of the challenges to conventional character coding are indicated. Conventional codings for these data include:

> tail colour: red (0); blue (1)
> tail colour: red (0); blue (1); purple (2)
> tail shape: straight (0) twisted (1)

Non-conventional character coding approaches are classified as nominal variable (presence/absence), unspecified homologue, composite, ratio, logically-related, conjunction, unifying, inapplicable data, positional and mixed codings.

Challenges to conventional character coding

Nominal variable coding

One challenge to the conventional approach, referred to as nominal variable coding by Pimentel and Riggins (1987) and as presence/absence coding by Pleijel (1995), codes the presence or absence of each attribute as a separate binary character. Bistate and multistate characters can be coded as nominal variables. Thus for a group of taxa which all have tails that are red or blue we have:

> red tail colour: present (0); absent (1)
> blue tail colour: present (0); absent (1)

and if taxa are included with purple tails

> red tail colour: present (0); absent (1)
> blue tail colour: present (0); absent (1)
> purple tail colour: present (0); absent (1)

as opposed to the conventional approach which codes

tail colour: red (0); blue (1)

or

tail colour: red (0); blue (1); purple (2)

Absence may mean that there is no homologue to presence or simply that no homologue is specified (Nelson, 1994). The homologue to presence is only revealed when there are two or more nominal variable characters. Detecting nominal variables is not therefore simply counting absence/presence characters; two or more characters are only considered to be nominal variables if they are related such that they can be conceptualized as a conventional character.

Unspecified homologue coding

Coding approaches which specify one state but do not specify the second are referred to as unspecified homologue coding. For a group of taxa which all have tails that are red or blue, or for a group with red, blue and purple tails, we have:

tail colour: red (0) otherwise (1)

As noted above, the distinction between nominal variable coding and unspecified homologue coding is not easily made. For the 'tail colour: red (0) otherwise (1)' example it is not apparent whether a number of different tail colours are represented by 'otherwise' or whether the character is logically equivalent to one of a pair of nominal variable characters 'red tail colour: present (0) absent (1)'. When surveying character conceptualization from published lists of characters, a rule of thumb is required. In this case a tally of unspecified homologue characters is made by simply counting characters which included a state 'otherwise' or 'not as (0)'. Unspecified homologue coding may be cryptic if the character is descriptive and the states are 'present' or 'absent', or 'yes' or 'no', e.g. 'red tail colour: present (0) absent (1)'.

Composite coding

Composite coding (Wilkinson, 1995) describes the representation of the observed combinations of multiple features of a complex structure as a single character. A number of attributes (usually of the same structure or organ) which could be coded as independent characters are united. In the case of the taxa with red, blue, twisted and straight tails, if all red tails were twisted and all blue tails straight, a lack of biological independence might be hypothesised and an approach taken which would code one character:

tails: red and twisted (0); blue and straight (1)

whereas a conventional approach would code:

tail colour: red (0) blue (1)
tail shape: straight (0) twisted (1)

Ratio coding

Ratio coding refers to the use of ratios to describe the relation of two features. It is this relation that is considered to be systematically significant, e.g.

tail length: less than or approximately equal to wingspan (0)
greater than wingspan (1)

The use of a ratio to describe the shape of a structure or organ, e.g. 'wing shape: longer than broad (0) broader than long (1)', is a related but quite different form of ratio coding which is much more common in the literature than true ratio coding, where it is used to delimit states often for continuously varying characters. Two types of ratio coding are distinguished as ratio coding and single-structure ratio coding for the purpose of this survey. The former refers to characters which describe the relationships of two or more structures and the latter characters which are shape descriptors.

Logically related coding

If two or more characters are different descriptors of a single variable they are described here as logically related characters. Often logically related characters make two different delimitations of the same quantitative variation, e.g.

tail length: approximately 5 units (0) greater than 5 units (1)
tail length: up to 10 units (0) more than 10 units (1)

However, not all logically related characters describe quantitative variation.

Conjunction coding

Conjunction coding describes any coding which recognises two character states, and includes a third character state to account for organisms which show both states. Thus we might have:

tail colour: red (0) blue (1) red and blue (2)

Note that conjunction coding differs from unifying coding (described below) which is concerned with conceptualizing characters where taxa may be polymorphic, whereas conjunction coding refers to 'polymorphism' within an individual organism.

Unifying coding

Unifying coding describes a situation where two states which could be coded as distinct are lumped together. Characters including states which bracket a range of quantitative variation within a single state e.g. 'red or reddish-orange', quite distinct from a second state 'blue', were not considered to be significantly different from conventional character coding, and were not scored as unifying characters in this survey. Only when the 'or' is used to unite two distinct states is the character referred to as a unifying character, e.g.

tail colour: red or blue (0) purple (1)

Inapplicable data coding

Character coding is problematic when some taxa have one form of a structure, others have a different form and the remaining taxa lack the structure altogether (Platnick *et al.*, 1991; Bateman *et al.*, 1992; Maddison, 1993; Pleijel, 1995; Hawkins *et al.*, 1997). There are a number of approaches to coding this type of variation. The only approaches found in the present literature survey are as follows: either two characters are employed, the first describing the presence or absence of the structure and the second describing the forms of the structure and using a question mark to denote absence of that structure, or a single multistate treatment is used. Thus, if all tails were red or blue but some taxa lacked tails, we might have:

tails: present (0) absent (1)
tail colour: red (0) blue (1) absent (?)

while the multistate coding is:

tails: red (0) blue (1) absent (2)

In this survey neither coding is considered conventional, although Hawkins *et al.* (1997) have shown theoretical advantages of missing data approaches over multi-state ones. All inapplicable data characters are here scored as unconventional, either as inapplicable data (missing) or inapplicable data (multistate) characters.

The absence of the structure which otherwise shows variation may be cryptic, and is not always represented as a state 'absent'. For example:

tail feathers: one (0) two (1)
tail feathers: united (0) separated (1) not applicable (?)

while the multistate coding is:

tails: with two feathers, united (0) with two feathers, separated (1) with one feather (2)

In this case the second character is inapplicable to taxa with one tail feather, which are scored '?' for this character. Such characters were scored as inapplicable (cryptic, missing) or as inapplicable (cryptic, multistate) characters.

The character state 'absent' in what appears to be a multistate character conceptualization of a continuous variable can also be interpreted as a form of inapplicable character coding, e.g.

tail length: long (0) short (1) absent (2)

can also be scored as

tails: present (0) absent (1)
tail length: short (0) long (1) inapplicable (?)

Such characters were scored as inapplicable (continuous, missing) and inapplicable (continuous, multistate).

Positional coding

When a structure or an attribute is found in distinct topological positions (i.e. on different structures or as attributes of different structures) in different taxa, there are two approaches to coding. Either two characters are used to describe the presence or absence of the repeated structure or attribute at each position, or a single character is used which recognises each position as a state. For the tails example the distribution of the attribute colour might be coded:

red colour on tail: present (0) absent (1)
red colour on wing: present (0) absent (1)

or alternatively we have:

red colour: on tail (0) on wing (1)

The distribution of the structures feathers might be coded:

feathers on tail: present (0) absent (1)
feathers on wing: present (0) absent (1)

or alternatively we have:

feathers: on tail (0) on wing (1)

None of these codings are considered conventional. Whether referring to attributes or structures, these codings are referred to as positional (nominal) and positional (composite) codings, where positional (nominal) codings use separate characters to describe separate sites, and positional (composite) codings use single characters.

Mixed coding

A number of characters combine two or more non-conventional approaches to conceptualize variation. As Maddison (1993) notes, multistate coding can become unwieldy when a structure which is sometimes absent has a number of other attributes. To avoid reiteration of 'absent' as a character state, a combination of composite coding and inapplicable (multistate) coding may be employed. Thus to describe tail shape and tail colour, and to avoid scoring the absence of a tail twice, we might have:

> tails: present and red and straight (0) present and red and twisted (1) present and blue and straight (2) present and blue and twisted (3) present and purple and straight (4) present and purple and twisted (5) absent (6)

There are numerous other possible combinations of non-conventional character coding, for example a combination of composite and unspecified homologue coding may be employed, e.g.

> tails: red and twisted (0) otherwise (1)

Wherever a character combines two forms of non-conventional character coding it is scored twice.

Methods

The morphological matrices published in 1995 and 1996 in two journals (*American Journal of Botany* and *Systematic Botany* – not including *Systematic Botany Monographs*) and two conference proceedings volumes (*Advances in Legume Systematics*, Volume 7 (Crisp and Doyle, 1995) and *Monocotyledons: Systematics and Evolution*, Volumes 1 and 2 (Rudall *et al.*, 1995)) were examined. In total 34 papers presented morphological matrices; a total of 36 matrices comprising 1404 characters were included in the survey. All the papers surveyed are listed in Appendix 2.1 (since all contained non-conventional characters), and are cited in the references.

All non-conventional character codings were recorded, and the total numbers of non-conventional characters for each of the categories described in the previous section were scored. It was noted whether primary homology criteria were discussed; a record was kept of all relevant statements.

Literature review to determine primary homology assessment approaches was sometimes problematic. In some cases the discussion of methods clarified the approaches to primary homology assessment taken, but often this was not the case since primary homology assessment criteria were not discussed or were glossed over. In all cases it was necessary to attempt to deduce relevant information from the character lists and data matrices. There are problems with this. Lack of familiarity with many of the morphological structures meant that in many cases the relationships between characters and states were unclear. As Patterson and Johnson (1997) point out, only those who are intimately familiar with a group of organisms can fully evaluate a data matrix. Because only those cases which seemed unambiguously non-conventional without recourse to examination of specimens were scored, the frequency of non-conventional character coding may be underestimated.

Results

The results of the survey are presented in Table 2.1. All non-conventional characters are listed in Appendix 2.1. Relevant discussions of primary homology criteria are presented in Appendix 2.2. All the matrices examined included at least one character considered non-conventional. Table 2.1 shows that although the majority of characters in each matrix are conventional, significant deviation from conventional character coding is found in the literature. Unspecified homologue coding is far more frequently employed than nominal variable coding, and is particularly favoured by some botanists e.g. Grimes (1995). Composite coding, the subject of a study by Wilkinson (1995), is very rarely used in practice. Positional coding is also rare. Ratio, logically-related and conjunction codings are all found with greater frequency than nominal variable codings. However, most significant deviation from conventional coding relates to the inapplicable data situation. There is a strong preference in practice for the multistate coding approach for inapplicable data.

Discussion

The framework of homology testing

Non-conventional coding is considered here in the context of the wider framework that is used to make primary homology decisions (Fig. 2.1). According to this framework (drawn from the work of Patterson (1982), presented by Rieppel (1988) and discussed by de Pinna (1991)), variation is examined and primary homology statements are conceptualized following similarity and conjunction criteria. Patterson (1982) described the complement relation as the relation between structures that do not conform to the similarity criterion but pass the congruence test. He clearly stated that repeated structures (homonomy) fail the conjunction criterion but are useful in systematics, perhaps thereby suggesting that a less rigorous application of the criteria would yield systematically informative characters. However, in the following discussion it is assumed that the characters in a cladistic data matrix must conform to the similarity and conjunction criteria. Following the

Table 2.1 Results of literature survey

	Matrices	Characters
Total number examined	34	1404
Nominal variable coding*	5	11
Unspecified homologue coding	7	47
Composite coding	6	7
Ratio coding	10	18
Logically related coding	6	7
Conjunction coding	7	11
Unifying coding	9	11
Inapplicable data coding*	29	154
Positional coding*	2	4

Note: A more detailed breakdown of character coding for characters indicated * is given in Appendix 2.3.

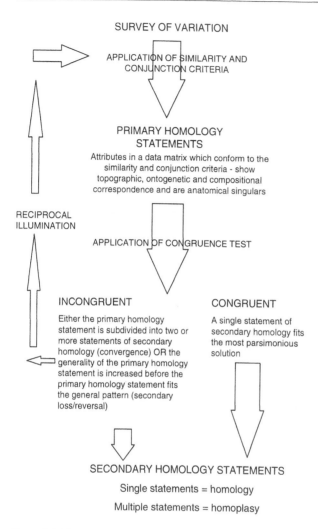

SURVEY OF VARIATION

APPLICATION OF SIMILARITY AND
CONJUNCTION CRITERIA

PRIMARY HOMOLOGY
STATEMENTS

Attributes in a data matrix which conform to the
similarity and conjunction criteria - show
topographic, ontogenetic and compositional
correspondence and are anatomical singulars

RECIPROCAL
ILLUMINATION

APPLICATION OF CONGRUENCE TEST

INCONGRUENT

Either the primary homology
statement is subdivided into two or
more statements of secondary
homology (convergence) OR the
generality of the primary homology
statement is increased before the
primary homology statement fits
the general pattern (secondary
loss/reversal)

CONGRUENT

A single statement of
secondary homology fits
the most parsimonious
solution

SECONDARY HOMOLOGY STATEMENTS

Single statements = homology

Multiple statements = homoplasy

Figure 2.1 The framework of homology testing. Rieppel (1988: 60) presented a schema summarizing the steps of phylogenetic reasoning. The framework presented here differs in that similarity is recognized as a criterion and not as a test, and conjunction is included with similarity as a criterion determining the form of a primary homology statement; topographic correspondence is not distinct from the similarity criterion which comprises topographic, ontogenetic and compositional correspondence; reciprocal illumination is included as a determinant of the primary homology statement.

application of the similarity and conjunction criteria, the congruence test is applied and primary homology statements are identified as single statements of secondary homology (congruence = homology) or multiple statements of secondary homology (incongruence = homoplasy). Where there are multiple statements of secondary homology, primary homology statements are re-examined through the process of reciprocal illumination. Elements of this scheme are considered here to show how non-conventional coding found in the literature is indicative either of failure to adopt the scheme or at least of different interpretations of some of its elements.

The similarity criterion

The most coherent discussion of the similarity criterion is that of Patterson (1982), whose work has led to the recognition of the role of topographic, compositional and ontogenetic correspondence in character conceptualization. Only one character clearly contravened the similarity criterion as it is widely interpreted: Smith and Doyle (1995) disregarded the ontogenetic correspondence argument (character 9, Nutlet wings: absent (0) derived from the bract (1) derived from the bracteole (2)). However, there were many examples which suggested that the similarity criterion is interpreted in different ways, for example if nominal variable (or unspecified homologue) and conventional character coding are compared. Where nominal variable coding is employed, the red tail colour of taxon A is similar (shows topographic, compositional and ontogenetic correspondence) to a red tail in B. The coding approach is a one-step approach: this is a red tail; which of the other taxa have identical red tails? In the conventional approach the red tail colour of taxon A is 'the same but different' to the blue tail colour of taxon B, but is identical to red tail colour in other taxa. The conventional approach is a two-step process (Brower and Schawaroch, 1996), comprising a topographic identity step (the redness in A and the blueness in B are both the colour of the tail) and a character state identity step (redness and blueness are distinguished, then scored for each terminal taxon). The conventional approach is explicit about the role of transformation in grouping; in contrast, following a nominal variable approach individual observations only become characters because transformation is implied between the specified and unspecified states. The central role of comparative biology is denied by those employing a nominal variable approach. The same criticisms can be made of unspecified homologue coding. In effect, two different similarity criteria are operating. Nominal variable and unspecified homologue coding are both one-step processes which are fundamentally different from conventional approaches. This study shows that conventional and one-step approaches are often found within a single matrix.

Different interpretations of the similarity criterion may be ascribed in part to theoretical confusion surrounding the role of transformation in primary homology assessment (Hawkins et al., 1997). However, even if there is consensus regarding the role of transformation, there are still special cases where the application of the similarity criterion is problematic. For example, positional coding is employed when a structure (or an attribute) is repeated. Repeated structures (or attributes of structures) show compositional and ontogenetic correspondence, but the different topographic placements of the structures (or attributes) may characterize different terminal taxa. In the case of a repeated structure (or attribute), topographical correspondence is difficult to interpret. In the case of inapplicable data coding, another difficulty in the application of the similarity criterion arises: should absence of a structure be considered similar to the presence of a form of a structure?

Application of the similarity criterion is tied up with notions of independence. The conjunction criterion also pertains to independence. Any re-evaluation of the similarity criterion must also address independence and conjunction.

Independence and conjunction

De Pinna (1991: 380) wrote that 'the decision whether any two or more attributes comprize a single transformation series or two or more independent series is one of the most basic, albeit still confusing, issues in systematics'. Similarity is interpreted here as indicating dependence among related character states – if structures conform to the similarity criterion, if they are 'the same but different', then one character should be recognised. Conjunction, on the other hand, indicates independence between unrelated character states. I believe that, by informing of dependence and independence relations, the two criteria indicate whether one or two characters should be recognised. Yet, as shown above, the similarity criterion may be interpreted in different ways. Interpretation of the conjunction argument is also confused.

Similarity and dependence are related as follows. If flowers are either red or blue, the presence of red in one taxon is equivalent to the absence of blue and the presence of blue is equivalent to the absence of red. Redness complements blueness; the two states constitute a whole with no other possible condition. The relationship between red and blue is a dependence relationship. Dependence can be characterized solely by the distribution of characters, two states constitute a whole with no other condition observed, but dependence is meaningless without the similarity criterion. Feathers and flowers could be considered as dependent: organisms which have flowers never have feathers and vice versa, but it would be a mistake to code the presence of flowers and feathers as states of one character. Dependent conditions must be treated as states of one character only if they conform to the similarity criterion.

The conjunction argument suggests a complementary and pragmatic definition of independence. In his discussion of the conjunction test, Patterson (1982: 38) simply stated that 'if two structures are supposed to be homologous, that hypothesis can be conclusively refuted by finding both structures in one organism'. When two structures are found in one organism they are independent.

Although seemingly very simple, the conjunction argument is a difficult one because conjunction can be interpreted in two ways. Firstly, the conjunction argument may be used to dictate the form of a primary homology statement, like the similarity criterion operating logically prior to the congruence test. In this case the coding 'tail colour: red (0) blue (1) red and blue (2)' should not be employed when redness and blueness co-occur in a single organism, since the criterion is failed. The occurrence of redness and blueness together informs us that the two are not homologous – thus they should not be encoded as putative homologues, i.e. as a single primary homology statement. An alternative coding should be used which does not include redness and blueness within a single character in the matrix.

However, conjunction also serves as an indicator of homoplasy, and is therefore placed with congruence and not with similarity. De Pinna (1991) explained how conjunction operates as a weak test or indication of homoplasy. The conjunction argument indicates a mistake in establishing primary homologies, but the mistake is only revealed in the context of the discovered scheme of relationships. Conjunction therefore operates as a weak test, one that need not be invoked, and does not directly address the issue of primary homology assessment.

My survey shows that a number of workers do not use the conjunction argument to inform primary homology assessment directly. That 11 characters were discovered which failed conjunction as a criterion suggests that if the conjunction criterion is employed in primary homology assessment, it is not used universally. Only Rudall and Cutler (1995: 160) discussed the conjunction argument ('Raphides are narrow needle-like crystals of calcium oxalate present in groups or bundles within idioblasts ... Raphides differ from the larger, solitary styloids, which may (in a few taxa, e.g. *Dracaena* and *Nolina*) coexist with raphides in the same plant, and must therefore be treated as a separate (i.e. non-homologous) character'). Such explicit use of the conjunction criterion is rare. The use of 'conjunction coding' in the literature demonstrates the need for a re-evaluation of the role of the conjunction argument.

I argue that conjunction coding should be avoided because the independent states are coded as one character. Independence/dependence arguments can be used to reject other forms of non-conventional coding. Nominal variable and logically-related coding introduce logical non-independence: the characters conceptualized fail a dependence criterion, since knowledge that a taxon has a particular character state for one character imposes a restriction upon the possible character states for another character (Wilkinson, 1995). Nominal variables and logically-related characters should be avoided because the characters are dependent. One might argue, on the basis of logical dependence, against the use of cladistic characters which invoke ratios. An independence argument is difficult to apply in other cases of non-conventional coding, such as inapplicable data coding or unifying coding: should hierarchically-related variation be conceptualized as two independent characters? Should some attributes be unified? Wilkinson (1995) distinguished between biological and logical independence, and discussed the implications of inferred biological non-independence and composite coding. The paucity of composite codings in the literature suggests that the majority of workers intuitively avoid invoking biological non-independence when conceptualizing characters.

Reciprocal illumination

The re-evaluation and modification of primary homology statements in order to minimize character conflict is referred to as reciprocal illumination (Siebert, 1992). There are three levels of re-evaluation if conventional character coding has been employed. Firstly, were the observations made correctly (has this taxon really got red tails)? Secondly, were the topographic/ontogenetic/compositional similarity decisions correct (are red and blue really the same but different)? Finally, were character state delimitation decisions correct (are redness and blueness really delimited in this way)? In the context of the present survey of non-conventional character coding, conflicting interpretations of reciprocal illumination are relevant because some approaches to character conceptualization limit the scope of reciprocal illumination. Other authors may arrive at non-conventional character coding approaches while attempting to minimize homoplasy through a process of reciprocal illumination. Grimes (1995) was explicit about this. He argued that a first analysis should be performed using a nominal variable matrix, and that following the first analysis characters should be recoded to reflect the transformational relationships discovered by congruence. For him, reciprocal illumination is not a process which permits re-eval-uation of topographic/ontogenetic/compositional identity decisions, rather these are determined *de novo*. While Grimes's (1995) approach reinterprets reciprocal illumi-nation, nominal variable coding denies one aspect of its operation: there is no 'same but different' to re-evaluate. Some authors consider this to be an advantage of the nominal variable approach; for example, Pleijel (1995: 312) avoids 'assumptions regarding character transformation that will never be tested in the analysis'.

Conclusions

This study supports the observations of authors such as Pleijel (1995: 309) who noted 'a large number of case studies, where different coding techniques are employed without discussion and applied in mixed ways within a single analysis'. Most workers adopt a conventional approach to character conceptualization, so that the character states of one character describe structures which are the same but different, yet all the matrices examined here include at least one non-conven-tional character. Approximately 16% of all characters are non-conventional. The breakdown in the application of the conventional character concept suggests that either guidelines are ignored or they are not interpreted in the same way.

The majority of studies of alternative coding approaches have evaluated nominal variable coding (e.g. Pimentel and Riggins, 1987; Cranston and Humphries, 1988; Pleijel, 1995; Wilkinson, 1995). The present survey shows that nominal variables are not among the most significant non-conventional coding approaches and that nominal variable coding has become a strawman in the face of other more wide-spread non-conventional coding practices. Only four of the 34 matrices were uniformly bistate, yet five of 11 nominal variable characters were part of uniformly bistate matrices, suggesting that often it is a preference for bistate coding which forces multi-condition variation into nominal variable coding. This situation might change following publication of Pleijel's (1995) paper which explicitly advocates the nominal variable approach. Although arguements such as Pleijel's (1995) promote a change from coding a predominantly conventional matrix with one or two nominal variable characters to employing a uniformly nominal variable matrix, any such change will be in the face of recent papers which have sought to clarify the reasoning behind the conventional approach (e.g. Brower and Schawaroch, 1996; Hawkins *et al.*, 1997).

Most significant deviation from conventional coding relates to the inapplicable data situation, where issues of similarity and independence are both problematic. Is absence of a structure similar to the presence of an attribute of a structure? Can hierarchically-related variation be considered independent? The prevalence of the inapplicable data problem shows that most studies (29 of 34 data matrices) must handle significant numbers of hierarchically-related characters. The debate surrounding the inapplicable data problem is very necessary, especially given the impact of different coding approaches on discovered topologies and character opti-mizations (Platnick *et al.*, 1991; Maddison, 1993; Hawkins, 1996; Hawkins *et al.*, 1997).

Character recognition is coloured by theory (Pogue and Mickevich, 1990). Given the confusion surrounding the framework which informs primary homology assessment, and confusion over the definition of the term 'character' (Hawkins *et al.*, 1997), it is not surprising that organismal variation is often conceptualized as characters and character states in different ways.

Studies showing the importance of character conceptualization on discovered topologies (e.g. Maddison, 1993; Hawkins *et al.*, 1997) underline the need for more rigorous evaluation of primary homology assessment. However, a call for a more rigorous evaluation and application of homology theory presumes that there can be a set of rules which can guide the objective representation of complex three- or four-dimensional organismal variation as a two-dimensional matrix. At present, different botanists are perceiving and defining characters in different ways; the discussion presented here shows that often inconsistencies can be ascribed to a failure of theory. However, many might argue that the construction of a morphological cladistic data matrix can only be a subjective process, guided by biological knowledge or insight, and that inconsistency is therefore to be expected. Primary homology assessment is subjective, imprecise and intuitive. To what extent must it remain so? I believe that the theoretical framework informing character conceptualization has not yet been fully explored, and that better guidelines are required. Until there is a clarification of theory inconsistency will remain.

Acknowledgements

I am very grateful to Robert Scotland, who invited me to contribute to this symposium of the Systematics Association. The groundwork for this study was completed as part of my doctoral studies at the Department of Plant Sciences, University of Oxford; I'm grateful to both Robert Scotland and Colin Hughes for their stimulating supervision of my doctoral project. I would like to thank the University of Cape Town for awarding the Smuts Fellowship, which allowed me to finish the manuscript. I also thank Peter Linder and Tony Verboom for comments, and Colin Hughes for reviewing the manuscript.

Appendix 2.1: Characters which are not coded 'conventionally'

Non-conventional characters are grouped by source (journal or volume) and author, and classified as follows:

- conventional, e.g. 'tail colour: red (0); blue (1)'
- nominal variable, e.g. 'red tail colour: present (0); absent (1)' and 'blue tail colour: present (0); absent (1)'
- unspecified homologue, e.g. 'tail colour: red (0) otherwise (1)'
- composite, e.g. 'tails: red and twisted (0); blue and straight (1)'
- ratio, e.g. 'tail length: less than or approximately equal to wingspan (0) greater than wingspan (1)'
- single-structure ratio, e.g. 'wing shape: longer than broad (0) broader than long (1)'
- logically related, e.g. 'tail length: approximately 5 units (0) greater than 5 units (1)' and 'tail length: up to 10 units (0) more than 10 units (1)'
- conjunction, e.g. 'tail colour: red (0) blue (1) red and blue (2)'

- unifying, e.g. 'tail colour: red or blue (0) purple (1)'*
- inapplicable data (missing), e.g. 'tails: present (0) absent (1)' and 'tail colour: red(0) blue (1) absent (?)'
- inapplicable data (multistate), e.g. 'tails: red (0) blue (1) absent (2)'
- inapplicable data (multistate, cryptic), e.g. 'tails: with two feathers, united (0) with two feathers, separated (1) with one feather (2)'
- inapplicable data (missing, cryptic), e.g. 'tail feathers one (0) two (1)' and 'tail feathers: united (0) separated (1)'
- inapplicable data (multistate, continuous), e.g. 'tails: absent (0) short (1) long (2)'
- inapplicable data (missing, continuous), e.g. 'tails: present (0) absent (1)' and 'tail length: long (0) short (1)'
- positional data (nominal), e.g. 'red colour on tail: present (0) absent (1)' and 'red colour on wing: present (0) absent (1)'
- positional data (composite), e.g. 'red colour: on tail (0) on wing (1)'.

'+' is used to indicate sets of nominal variable, logically related and inapplicable data (missing) characters; the classification is indicated only for the first member of each set.

American Journal of Botany

Gustafsson and Bremer (1995). 46 CHARACTERS
13 Calyx: actinomorphic (0) zygomorphic with odd sepal dorsal (1) zygomorphic with odd sepal ventral (2). INAPPLICABLE DATA (MULTISTATE, CRYPTIC)
14 Corolla: actinomorphic (0) bilabiate 3 + 2 (1) bilabiate 1 + 4 (2) bilabiate (1 + 1) + 3 (3). INAPPLICABLE DATA (MULTISTATE, CRYPTIC)
18 Petal lateral veins: absent (0) present (1). INAPPLICABLE DATA (MISSING + 19 AND 20)
19 Petal lateral veins: ending subapically (0) apically confluent (1).
20 Lateral veins: free (0) fused with adjacent lateral (1).

Smith and Doyle (1995). 38 CHARACTERS
9 Nutlet wings: absent (0) derived from the bract (1) derived from the bracteole (2). FAILS THE SIMILARITY CRITERION; INAPPLICABLE DATA (MULTISTATE)
15 Stigma: subglobose and nondivided (0) commissural and divided along the plane of carpel suture (1) carinal and divided along the centre line of the carpels (2).
INAPPLICABLE DATA (MULTISTATE, CRYPTIC)

Luckow and Hopkins (1995). 52 CHARACTERS
30 Anther gland: present (0) absent (1). INAPPLICABLE DATA (MISSING + 31)

---

*One character of the form 'tail colour: red or blue (0) blue (1)' (Linder and Kellogg, 1995, character 2) was scored as a unifying character, although it differs in form from unifying characters as discussed here.

31 Attachment of anther gland: sessile (0) short-stipitate (1) long-stipitate (2).
35 Vestiture on valves: lacking (0) velutinuous (1) scurfy (2). INAPPLICABLE DATA (MULTISTATE)
36 Fruit dehiscence: indehiscent (0) dehiscent by one suture (1) dehiscent by two sutures (2). INAPPLICABLE DATA (MULTISTATE, CRYPTIC)
49 Exine sculpturing: reticulate (0) verrucate (1). INAPPLICABLE DATA (MISSING + 50)
50 Exine reticulations: not raised (0) raised (1).

Axelius (1996). 41 CHARACTERS
5 Stellate hairs outside corolla: absent (0) present (1). POSITIONAL DATA (NOMINAL, +6)
6 Stellate hairs outside calyx: absent (0) present (1).
16 Calyx tube: not enlarging with fruit maturity (0) enlarging in pace with fruit (1) enlarging in advance of fruit (2) INAPPLICABLE DATA (MULTISTATE, CRYPTIC, CONTINUOUS); RATIO
22 Lobes: more or less absent (0) small, much shorter than tube (1) long, longer top slightly shorter than tube (2). INAPPLICABLE DATA (MULTISTATE, CONTINUOUS); RATIO
24 Second colour: absent (0) green to yellowish (1) blue to purple (2). INAPPLICABLE DATA (MULTISTATE)
25 Form of marking: absent (0) five ± rounded spots at each midvein (1) small spots forming five areas along the midvein in the throat (2). POSITIONAL DATA (COMPOSITE); INAPPLICABLE DATA (MULTISTATE)
28 Hairiness in throat: absent (0) situated around five distinct nectar holes (1) most abundant behind the stamens (2) most abundant between the stamens, in distinct spots between the stamens (3). POSITIONAL DATA (COMPOSITE); INAPPLICABLE DATA (MULTISTATE)
29 Stamens: much shorter than style (0) ± same length to longer than style (1). RATIO
34 Fruit: a capsule (0) a thin-walled berry, at first somewhat juicy, but later dry ± leathery (1) a juicy berry (2) a thick-walled dry berry (3). INAPPLICABLE DATA (MULTISTATE, CRYPTIC)
40 Holes in lower periclinal wall: none (0) small pores (1) holes (2). INAPPLICABLE DATA (MULTISTATE)

Ronse Decraene *et al.* (1996). 42 CHARACTERS
2 Leaves: simple (0) compound with odd number of leaflets (1) compound with even number of leaflets (2). INAPPLICABLE DATA (MULTISTATE, CRYPTIC)
20 Nectaries: absent (0) receptacular, disc-like (1) receptacular, not disc-like (1). INAPPLICABLE DATA (MULTISTATE); UNSPECIFIED HOMOLOGUE
40 Rays: absent (0) uniserate (1–3 cells wide) (1) multiserate (>3 cells wide) (2). INAPPLICABLE DATA (MULTISTATE)

Lane and Hartman (1996). 17 CHARACTERS
6 Leaf glandularity: none (0) punctuate (1) stipitate (2). INAPPLICABLE DATA (MULTISTATE)

10 Phyllary apex glands: none (0) sessile (1) punctuate (2). INAPPLICABLE DATA (MULTISTATE)
13 Ray and disc corolla tube glands: sessile or capitate (0) none (1). UNIFYING

Hufford (1996). 15 CHARACTERS
4 Female inflorescences: racememes or axillary flowers (0) strictly axillary flowers.
CONJUNCTION
5 Pedicel length for female flowers: >10 mm (0) 0.5–7 mm (1) absent (2). INAPPLICABLE DATA (MULTISTATE, CONTINUOUS)
13 Fruit dehiscence: ventral (0) entire margin (1) indehiscent (2) INAPPLICABLE DATA (MULTISTATE, CRYPTIC)
14 Aril size: absent (0) extending less than a third of overall seed length (1) greater than half the length of the seed (2). INAPPLICABLE DATA (MULTISTATE); RATIO

Systematic Botany

Graham (1995). 25 CHARACTERS
8 Floral tube shape: as wide as long (0) longer than wide (1). SINGLE-STRUCTURE RATIO
10 Stamen/sepal ratio: 2/1 (0) 1/1 (1) 3+/1 (2). RATIO
11 Carpel/perianth ratio: 1/1 (0) 1/1 (1). RATIO
14 Calyx lobes: erect or none (0) spreading or reflexed (1). UNIFYING
20 Nectary position: basal (0) non-basal (1) absent (2). INAPPLICABLE DATA (MULTISTATE)
23 Seed wing: absent (0) present/unilateral (1) present/encircling (2). INAPPLICABLE DATA (MULTISTATE)
24 Seed coat hairs: absent (0) present, straight (1) present, spiral (2). INAPPLICABLE DATA (MULTISTATE)

Guala (1995). 13 CHARACTERS (ALL BISTATE)
6 Plants: fire adapted (0) not (1). COMPOSITE

Lutzoni and Brodo (1995). 61 CHARACTERS
6 Dark excipulum probrium: absent (0) continuous below the subhymenium (1) present in the apothecial margin only (2). INAPPLICABLE DATA (MULTISTATE)
12 Hymenial reaction to HNO_3: negative (0) positive violaceous pink (1) positive orange/yellow. INAPPLICABLE DATA (MULTISTATE)

Weller et al. (1995). 43 CHARACTERS
12 Reduction in leaf venation: leaves 3- or more-nerved (0) leaves 1-nerved (1). LOGICALLY RELATED (+ 13)
13 Increase in leaf venation: leaves 1- or 3-nerved (0) 5-nerved (1) 7-nerved (2).
15 Pubescence on leaf margin: no pubescence (0) thin hairs present (1) hooked hairs present (2). INAPPLICABLE DATA (MULTISTATE)

De Luna (1995). 41 CHARACTERS
1 Spores and protonema: spores unicellular and protenema exosporic filamentous
(0) spores unicellular but protonema development exosporic and globular (1) spores
multicellular. INAPPLICABLE DATA (MULTISTATE, CRYPTIC), COMPOSITE
8 Pseudoparaphyllia: absent (0) filamentous, uni- or bi-serate (1) foliose, very
broad at base (2). INAPPLICABLE DATA (MULTISTATE, MISSING + 9)
9 Surface of pseudoparaphyllia: does not apply (?) surface smooth (0) surface
papilose (1).
10 Leaf costa: single (0) absent (1) double (2). INAPPLICABLE DATA
(MULTISTATE)
23 Perichaetial paraphyses: unknown (?) present, short, hyline (0) present, long,
becoming yellowish after fertilization (1) absent (2). INAPPLICABLE DATA
(MULTISTATE)
36 Differentiation of exostome: uncertain (?) present, 16 teeth (0) present, teeth
fused into 8 pairs (1) absent (2). INAPPLICABLE DATA (MULTISTATE,
MISSING + 37 AND 39)
37 Shape of exostome tooth: does not apply (?) teeth long, slenderly lanceolate
(1) teeth short, broadly accuminate (2) teeth short, truncated (2).
39 Movement of peristome: does not apply (?) teeth xerocastique (1) teeth
hydrocastique (2).
41 Segment of the endostome: tall well differentiated (0) short poorly developed
(1) not present (2). INAPPLICABLE DATA (MULTISTATE, CONTINUOUS)

Goldblatt and Manning (1995) 25 CHARACTERS
23 Flowers: odourless (0) weakly fragrant (1) strongly fragrant (2).
INAPPLICABLE DATA (MULTISTATE, CRYPTIC, CONTINUOUS)

Karis (1995) 52 CHARACTERS
4 Leaves: simple (0) dissected into narrow lobes (1) dissected and secondarily
filiform (2).
INAPPLICABLE DATA (MULTISTATE, CRYPTIC)
13 Specialized fruit enclosing receptacles: absent (0) present (1). INAPPLICABLE
DATA (MISSING + 14, 15 AND 16)
14 Specialized fruit enclosing receptacle spine tips: straight (0) hooked (1).
15 Specialized fruit enclosing receptacle appendages: evenly distributed (0)
confined to the middle of the receptacle (1)
16 Specialized fruit enclosing receptacles: smooth (0) tuberculate to spiny (1)
winged (2).
20 Corollas: with multiseptate helianthian hairs (0) corollas glabrous or
glandular (1).
INAPPLICABLE DATA (MISSING + 21), UNIFYING
21 Multiseptate helianthian hairs: with an acute top cell (0) with an obtuse top
cell (1).

Lammers (1996) 21 CHARACTERS
12 Hypanthium shape: longer than broad (0) broader than long (1). SINGLE
STRUCTURE RATIO

Liede (1996) 41 CHARACTERS
16 Corolline corona: present (0) absent (1). INAPPLICABLE DATA
(MISSING + 21)
17 Corolline corona: glabrous (0) hairy (1) inapplicable (?)
18 Anther twisting: none (0) moderately (1) largely (2). INAPPLICABLE DATA
(MULTISTATE, CONTINUOUS)
19 Anther abortion: none (0) 1–3 aborted stamens (1) only 1 fertile stamen (2).
INAPPLICABLE DATA (MULTISTATE, CONTINUOUS)
25 Carpel ventral traces: free (0) fused at origin (1) fused throughout (2).
INAPPLICABLE DATA (MULTISTATE, CONTINUOUS)

Ambruster (1996) 63 CHARACTERS
12 Distal pistillate involucre bractlets: present (0) absent (1). INAPPLICABLE
DATA (MISSING + 13)
13 Form of distal pistillate involucre bractlets: stipuliform (0) involucral (1)
inapplicable (?).
17 Connation of staminate involucellar bracts: free (0) connate or bilabiate (1)
fused at base only (2). INAPPLICABLE DATA (MULTISTATE, CONTINUOUS)
28 Relative petiole length: >1/4 limb (0) <1/10 limb (1). RATIO
59 Sepal length at anthesis: approx. as long as ovary (0) >twice as long as
ovary (1).
RATIO

Monocotyledons: Systematics and Evolution, volumes 1 and 2

Rudall and Cutler (1995) 36 CHARACTERS
13 Stamens: six (0); three, lacking outer whorl (1); three, lacking inner whorl
(2).
CONJUNCTION
26 Embryo sac type: *Polygonum* (0) other (1). UNSPECIFIED HOMOLOGUE
27 Antipodal: normal or ephemeral (0) giant (1) UNIFYING

Goldblatt (1995) 23 CHARACTERS
5 Calcium oxylate present: as raphides (0) cuboidal crystals (1) raphides and
styloids (2) styloids (3) crystal sand (4) none (5). INAPPLICABLE DATA
(MULTISTATE), CONJUNCTION
21 Perianth whorls: both calycine (0) petaloid (1) outer calycine, inner petaloid
(2).
CONJUNCTION
22 Epicuticular waxes: lacking or of the non-oriented type (0) epicuticular waxes
of the *Convallaria*-type (orientation parallel) (1). UNIFYING
23 Stamens: six in two whorls (0) inner whorl lacking (1) at least posterior
three stamens lacking. CONJUNCTION

Kress (1995) 36 CHARACTERS (ALL BISTATE)
7 Internal silica bodies hat shaped: present (0) absent (1). NOMINAL
VARIABLE (+ 8 AND 9)

8 Internal silica bodies trough shaped: present (0) absent (1).
9 Internal silica bodies druse shaped: present (0) absent (1).
30 Silica bodies: absent (0) present (1). INAPPLICABLE DATA (MISSING +7, 8 AND 9)

Linder and Kellogg (1995) 47 CHARACTERS
1 Leaf insertion: distichous (0) spiral or tristichous (1). UNIFYING, LOGICALLY RELATED (+2)
2 Leaf insertion: spiral or tristichous (0) tristichous (1). UNIFYING*
7 Root hair cells: like other epidermal cells (0) shorter than in other epidermal cells, differentiated at the apex and with a denser cytoplasmic content (1) root hairs absent (2).
INAPPLICABLE DATA (MULTISTATE)
8 Sieve tube plastids: with protein bodies and starch grains (0) with only protein bodies (1) with only protein bodies, but these with fragments (2).
CONJUNCTION
12 Chorenchyma of culms and leaves: isodiametrical or lobed (0) with peg cells (1).
UNIFYING
41 Endosperm formation: nuclear (0) helobial (1). INAPPLICABLE DATA (MISSING + 44, CRYPTIC)
44 Helobial endosperm: with the chalazal chamber not forming cells (0) with chalazal chamber with 4–16 cells

Simpson (1995) 35 CHARACTERS
3 Pseudopetiole: absent (0) *Mapania* type (1) *Oreobolus* type (2).
INAPPLICABLE DATA (MULTISTATE)
9 Silica bodies: usually absent (0) cone-shaped (1) spheroidal (2).
INAPPLICABLE DATA (MULTISTATE)
Kellogg and Linder (1995) 28 CHARACTERS
15 Starch granules: many large (0) absent or small (1). UNIFYING
24 Microhairs: bicellular (0) multicellular (1) absent (2). INAPPLICABLE DATA (MULTISTATE)

Stevenson and Loconte (1995) 101 CHARACTERS
4 Lateral root origination: opposite xylem (0) opposite phloem (1) both (2).
CONJUNCTION
10 Sieve tubes plastids: starch only (0) P11c (1) P11cs (2) P11c′(3).
CONJUNCTION
18 Petiole: absent (0) *Dioscorea*-type (1) *Alisma*-type (2) *Bambusa*-type (3).
INAPPLICABLE DATA (MULTISTATE)
28 Epicuticular waxes: absent (0) *Strelizia*-type (1) *Convallaria*-type (2).
INAPPLICABLE DATA (MULTISTATE)
29 Stomates: anomocytic (0) paracytic (1) tetracytic (2) absent (3).
INAPPLICABLE DATA (MULTISTATE)
30 Leaf air canals: absent (0) random (1) one arc (2) two arcs (3) septate (4).
INAPPLICABLE DATA (MULTISTATE)

33 Perianth type: petaloid differentiated (0) sepaloid (1) peteloid undifferentiated (2) absent (3). INAPPLICABLE DATA (MULTISTATE)
39 Number of stamens: one (0) two (1) three from both whorls (2) four (3) five (4) six (5) many (6) three from inner whorl (7) three from outer whorl (8). CONJUNCTION
53 Number of pollen apertures: 0 (0) one (1) two (2) three (3) poly (4) INAPPLICABLE DATA (MULTISTATE, CONTINUOUS), INAPPLICABLE DATA (MISSING + 54)
54 Pollen aperture margin: non-annulate (0) annulate (1)
55 Pollen sculpture: reticulate (0) echinate (1) psilate (2) clavate (3) microreticulate (4) scrobiculate (5) scabrate (6) striate (7) exine (8) absent (9) INAPPLICABLE DATA (MULTISTATE)
59 Septal nectaries: absent (0) internal (1) external (2). INAPPLICABLE DATA (MULTISTATE)
61 Embryostega: absent (0) internal (1) external (2). INAPPLICABLE DATA (MULTISTATE)
62 Seed storage tissue: absent (0) endosperm (1) peristerm (2) chalazosperm (3). INAPPLICABLE DATA (MULTISTATE)
86 Cotyledon sheath: absent (0) open (1) closed (2) INAPPLICABLE DATA (MULTISTATE)
87 Cotyledon ligule: absent (0) open (1) closed (2) INAPPLICABLE DATA (MULTISTATE)

Uhl *et al.* (1995) 41 CHARACTERS
15 Peduncular bracts: absent (0) present (1). INAPPLICABLE DATA (MISSING + 15)
16 Peduncular bracts: several (0) one (1).
19 Flowers: bisexual only present (0) bisexual and unisexual present (1) unisexual only (2). CONJUNCTION, INAPPLICABLE DATA (MISSING + 20, CRYPTIC)
20 Taxa with unisexual flowers monoecious (0) dioecious (1) polygamo-dioecious (2) polygamo-dioecious (3).
27 Carpels: distinct (0) connate (1). INAPPLICABLE DATA (MISSING + 29, CRYPTIC)
29 Carpels: connate throughout (0) connate by styles only (1).
33 Endocarp: lacking pores (0) with apical pore (1) with three clearly defined pores (2). INAPPLICABLE DATA (MULTISTATE)

Cox *et al.* (1995) 18 CHARACTERS (ALL BISTATE)
All 18 characters were presented in the form 'attribute: no (0) yes (1)', where the attribute described was considered to be the apomorphic state. All 18 characters were therefore considered to employ unspecified homologue coding.
13 Root vessels with both scaliform and simple perforation plates: no (0) yes (1) CONJUNCTION

Advances in Legume Systematics

Lewis and Schrire (1995) 24 CHARACTERS

10 Standard claw: absent or simple (0) folded or grooved (1) forming an inrolled tube (2). UNIFYING, INAPPLICABLE DATA (MULTISTATE)

13 Stamen filaments: glabrous (0); with a small tuft of hairs at base (1) hairy for at least half to two-thirds of length (2). INAPPLICABLE DATA (MULTISTATE, CRYPTIC)

14 Stamen hairs: absent (0) rusty brown or reddish (1) white or hyaline (2). INAPPLICABLE DATA (MULTISTATE)

15 Stamen length: about equalling petals or shorter (0) exerted from corolla (1) RATIO

18 Stigma rim: absent (0) present but unfringed (1) totally fringed (2) partially fringed (3). INAPPLICABLE DATA (MULTISTATE)

23 Sepal hairs (excluding margin): present (0) absent (1) INAPPLICABLE DATA (MISSING +24)

24 Sepal hair colour: rusty brown (0) hyaline or absent (1)

Luckow (1995) 24 CHARACTERS

2 Anther glands: absent (0) stipitate, round (1) modified stipitate (2) appendiculate (3). INAPPLICABLE DATA (MULTISTATE)

6 Leaf nectary: cup-shaped, stipitate (0) cup-shaped, sessile (1) club-shaped (2) INAPPLICABLE DATA (MULTISTATE, CRYPTIC)

9 Inflorescence: spike (0) compressed spike (1) head (2). INAPPLICABLE DATA (MULTISTATE, CRYPTIC)

16 Stigma: porate (0) narrow funnelform (1) broad funnelform (2). INAPPLICABLE DATA (MULTISTATE, CRYPTIC)

20 Pollen: calymmate (0) acalymmate tetrads (1) acalymmate monads (2). INAPPLICABLE DATA (MULTISTATE, CRYPTIC)

21 Fruits: woody, indehiscent (0) coriaceous, dehiscent through one suture (1) coriaceous, indehiscent (2) coriaceous, inertly dehiscent through two sutures (3) coriaceous, elastically dehiscent through two sutures. COMPOSITE, INAPPLICABLE DATA (MULTISTATE, CRYPTIC)

Chappill and Maslin (1995) 73 CHARACTERS

13 Nerve anastomoses: absent (0) few (1) numerous (2). INAPPLICABLE DATA (MULTISTATE, CONTINUOUS)

15 Inflorescence: rachis axis absent (0) short, simple rachis axis (1) long simple rachis axis (2) branched raceme axis (3). COMPOSITE, INAPPLICABLE DATA (MULTISTATE, CONTINUOUS)

18 Basal peduncle bracts: absent (0) 1 or 2 (1) 3 or more (2). INAPPLICABLE DATA (MULTISTATE, CONTINUOUS)

22 Stamen fusion: free (0) irregularly connate (1) substantially fused (2). INAPPLICABLE DATA (MULTISTATE, CRYPTIC)

35 Cotyledon base shape: distinctly auriculate (0) slightly auriculate (1) not auriculate (2). INAPPLICABLE DATA (MULTISTATE, CRYPTIC, CONTINUOUS); UNSPECIFIED HOMOLOGUE

67 Polyads per anther: more than 8 (0) 8 (1) LOGICALLY RELATED (+68)

68 Polyad number: 4 (0) 8 (1) 12 (2) 16 (3)

Grimes (1995) 75 CHARACTERS

3 Meristems: auxotelic (0) anauxotelic (1) anauxotelic after one repeating growth unit (2). INAPPLICABLE DATA (MULTISTATE, CRYPTIC)

4 Vegetative growth: vegetative growth proleptic from the axils of persistent leaves (0) some modification of proleptic growth. Variations of proleptic branching occur (compare 5, 6, 7 and 8) and these are scored as simple binary characters. LOGICALLY RELATED (+ 5 AND 6)

5 Vegetative growth: vegetative growth proleptic from the axils of persistent leaves (0) vegetative growth from sylleptic and proleptic growth (1)

6 Vegetative growth: vegetative growth proleptic from the axils of persistent leaves (0) vegetative growth from proleptic buds arising after abscission of subtending leaf. Similar to 4 and 5, but in 4 proleptic buds arise while leaf is persistent.

7 Vegetative branches: all long shoots (0) some vegetative short shoots (1). INAPPLICABLE DATA (MISSING +9)

8 Shoots: not as 1 (0) brachyblasts form sylleptically, and are both vegetative and reproductive (1). UNSPECIFIED HOMOLOGUE

9 Short-shoots: not as 1 (0) inflorescences form sylleptically, then strictly reproductive short-shoots (1). UNSPECIFIED HOMOLOGUE

10 Leaf heterochrony: none (0) spatial (1) INAPPLICABLE DATA (MISSING + 11, 12 AND 13)

11 Leaf heterochrony: not as 1 (0) expressed by suppression of leaves at the proximal most, older nodes (1). UNSPECIFIED HOMOLOGUE

12 Leaf heterochrony: not as 1 (0) expressed by suppression of leaves at the distal most, youngest nodes. UNSPECIFIED HOMOLOGUE

13 Leaf heterochrony: not as 1 (0) expressed by delayed development of leaves subtending inflorescences (1). UNSPECIFIED HOMOLOGUE

14 Stipules: present (0) absent (1). INAPPLICABLE DATA (MISSING + 15 AND 16)

15 Stipules: persistent (0) deciduous (1)

16 Stipules: foliaceous (0) modified into spines (1)

25 Inflorescences: not as 1 (0) in part of annual, axillary, strictly reproductive, deciduous branch systems (1). UNSPECIFIED HOMOLOGUE

26 Inflorescences: not as 1 (0) peduncles becoming lignescent and modified into spines (1). UNSPECIFIED HOMOLOGUE

38 Pollen: not as 1 (0) with the occurrence of two supplementary sets of pores faced by fours (1). UNSPECIFIED HOMOLOGUE.

42 Pollen: not as 1 (0) inner-outer polyad dissymmetry (1). UNSPECIFIED HOMOLOGUE

53 Fruit: not dehiscent as 1 (0) elastically dehiscent from the apex, the valves not contorting (1). COMPOSITE, UNSPECIFIED HOMOLOGUE

55 Fruit: not as 1 (0) fruit follicular, narrow, septiferous and resinous inside (1). COMPOSITE, UNSPECIFIED HOMOLOGUE

57 Fruit: nonseptate (0) septate, the endocarp in discrete packets around the seeds, and not forming a continuous layer on the inside of the valve (1). INAPPLICABLE DATA (MISSING + 58, CRYPTIC)

58 Fruit: nonseptate (0) septate, but the septa separating at dehiscence (1).

60 Fruit: not as 1(0) valves of the pods external red and fleshy (1).
UNSPECIFIED HOMOLOGUE
62 Resinoid tissue: lacking (0) present, distributed nearly all around the seed
coat (1). POSITIONAL (NOMINAL) + 63
63 Resinoid tissue: lacking (1) present, but distributed only near hillum (2).
68 Seed coat: not as 1 (0) translucent in part, white in part (1). UNSPECIFIED
HOMOLOGUE
72 Confluent parenchyma: absent (0) present (1) abundant, paratrachel banded
parenchyma (2). INAPPLICABLE DATA (MULTISTATE)
74 Crystal-bearing fibres: absent (0) present in single chains (1) as biseriate
crystal chains (2). INAPPLICABLE DATA (MULTISTATE)

Herendeen (1995) 29 CHARACTERS
1 Leaves: multifoliate (0) simple (1). INAPPLICABLE DATA (MISSING +2,
CRYPTIC)
2 Leaflets: opposite (0) alternate (1).
4 Leaflets punctuate: absent (0) present (1). UNSPECIFIED HOMOLOGUE
12 Calyx lobes: imbricate (0) valvate (1) calyx entire (2). INAPPLICABLE
DATA (MULTISTATE, CRYPTIC)
14 Petal number: five (0) one (1) zero (2) six (3). INAPPLICABLE DATA
(MULTISTATE, CONTINUOUS)
16 Adaxial petal width: same width as other petals (0) slightly broader (1)
broad standard petal (2) RATIO
21 Long lanceolate anthers: absent (0) present (1). UNSPECIFIED HOMO-
LOGUE

Schrire (1995) 57 CHARACTERS
3 Biramous hairs: absent (0) present (1). INAPPLICABLE DATA (MISSING + 4)
4 Biramous hairs (arm length): equal (0) very unequal (1). RATIO
8 Pearl bodies (lower leaflet surface): absent (0) clavate (1) discoid (2).
INAPPLICABLE DATA (MULTISTATE)
11 Phylolodinous petioles; absent (0) present (1). UNSPECIFIED HOMOLOGUE
13 Leaflet margins (dentate): absent (0) present (1). UNSPECIFIED
HOMOLOGUE
14 Leaflets (involute margins): absent (0) present (1). UNSPECIFIED
HOMOLOGUE
15 Leaflets (coriaceous): absent (0) present (1). NOMINAL VARIABLE (+ 16)
16 Leaflets (sclerophyllous): absent (0) present (1).
19 Bracts (trilobed): absent (0) present (1). UNSPECIFIED HOMOLOGUE
22 Peduncle (filiform/capillary, 1-few-flowered): absent (0) present (1).
COMPOSITE, UNSPECIFIED HOMOLOGUE
25 Calyx (length of lobes in proportion to the tube): shorter to ± equalling the
tube (0) longer than the tube (1). RATIO, LOGICALLY RELATED + 26
26 Calyx (length of lobes in proportion to the tube): up to + twice as long as
the tube (0) twice to six times as long as the tube (1). RATIO
27 Calyx (scarious lobes): absent (0) present (1). INAPPLICABLE DATA
(MISSING + 28)

28 Calyx (enlarged scarious lobes): absent (0) present (1). UNSPECIFIED HOMOLOGUE
30 Standard (dorsal surface): hairy (0) glabrous (1). INAPPLICABLE DATA (MISSING + 31)
31 Standard (dorsal surface hair colour): hyaline (0) dark brown (1).
38 Stamens (fusion): free (0) partially fused (1) fused (2). INAPPLICABLE DATA (MULTISTATE, CRYPTIC, CONTINUOUS)
42 Ovary (length in proportion to style length): long ovary/short style (0) short ovary/long style (1). RATIO.
46 Pod (reflexed): absent (0) present (1). NOMINAL VARIABLE (+ 47)
47 Pod (erect): absent (0) present (1).

Crisp and Weston (1995) 52 CHARACTERS (ALL BISTATE)
1 Chromosomes 1, x = :9 (0) 6–8 (1). LOGICALLY RELATED (+2)
2 Chromosomes 2, x = :8–9 (0) 6–7 (1).
17 Stipules: conspicuous (0) reduced to absent (1). LOGICALLY RELATED (+ 18)
18 Stipules: none or normal (0) filiform to subulate (1).
22 Bracteolar node articulation: absent (0) present (1)
23 Bracteoles: absent (0) present (1). INAPPLICABLE DATA (MISSING + 22)
28 Petals: otherwise (0) red and/or green (1). UNSPECIFIED HOMOLOGUE
30 Standard purple veins abaxially: absent (0) present (1). UNSPECIFIED HOMOLOGUE
31 Keel/wings length: 1 (0) <1 (1). RATIO
42 Pod: otherwise (0) small and ovoid or ellipsoid (1). UNSPECIFIED HOMOLOGUE
43 False dissepiment lower suture: absent (0) present (1). UNSPECIFIED HOMOLOGUE
44 Testa reticulate rugosity: absent (0) present (1). NOMINAL VARIABLE (+ 45)
45 Testa irregular rugosity: absent (0) present (1).
49 Aril: present (0) absent or rim only (1). UNIFYING, INAPPLICABLE DATA (MISSING + 50, 51 AND 52)
50 Aril: continuous (0) interrupted at micropyle (1).
51 Aril: otherwise (0) channelled and papillate (1). UNSPECIFIED HOMO-LOGUE
52 Aril: not lobed (0) lobed (1).

Van Wyk and Schutte (1995) MATRIX I, 18 CHARACTERS; MATRIX II, 30 CHARACTERS; MATRIX III, 21 CHARACTERS
Matrix I, 6 Stamens: free (0) fused into an open tube (1) fused into a closed tube (2). INAPPLICABLE DATA (MULTISTATE, CRYPTIC, CONTINUOUS)
Matrix I, 8 Carinal anther size: similar to basifixed anthers (0) intermediate or similar to dorsifixed anthers (1). RATIO
Matrix II, 4 Petiole: present (0) reduced to a pulvinus only (1) absent (2). INAPPLICABLE DATA (MULTISTATE, CONTINUOUS)
Matrix II, 18 Stamen fusion: totally free (0) slightly fused (1) fused higher up (2). INAPPLICABLE DATA (MULTISTATE, CRYPTIC, CONTINUOUS)

Appendix 2.2: Discussion of character conceptualization

Discussion of character conceptualization is grouped by source (journal or volume) and author.

American Journal of Botany

Gustafsson and Bremer (1995) 'Unknown or inapplicable characters are coded with a question mark.'

Luckow and Hopkins (1995) 'Coding of pollen characters is rather complex and deserves comment. There are apparently four basic types of exine in *Parkia*: reticulate, raised reticulate, verrucate and fossulate. The exines found in grains of *P. velutina* Benoist., *P. bahiae* H.C. Hopkins and some of *P. multijuga* seem at first glance to be a modification of the verrucate type, but closer inspection revealed that they are really a modification of the reticulate type. Although it is clear that all four types are closely related, it is not possible to determine a transformation series between them all. Coding exine as four non-additive states would fail to distinguish the homology between reticulate and raised reticulate exines. In such a situation one can either code the character using step matrices (Maddison and Maddison, 1992) or negative weighting (Nixon, unpublished data) or break the character into more than one state, as was done here. The latter introduces many missing values as inapplicable, (e.g. all the non-reticulate taxa are scored as 'missing' for the character reticulations raised) but the analysis is faster than with step matrices and presents no problems as long as the resultant tree does not contain impossible optimisations (e.g. non-reticulate exines have raised reticulations).'

Systematic Botany

Guala (1995) 'Although well over 100 characters were examined and recorded for the three species (Guala, 1992) only 13 were chosen for the cladistic analysis. These were the only ones for which a large sample size was available, that were logically and justifiably divisible into two states, and that were entirely discrete in the distribution of those states between OTUs. In the case of a three taxon ingroup, non-bimodal characters would simply be phylogenetically informative unless they were ordered. No justification for ordering was found in any of the characters. My logic in the division of characters into states follows Stevens (1991). Both Chappill (1989) and Pimentel and Riggins (1987) argued for the use of qualitative over quantitative states, and I have tried to define character states in a qualitative way.' Character 6 represents a suite of characters.

Lutzoni and Brodo (1995) 'All characters used in previous studies on this complex were scored for [...] 20 individuals, for a total of 92 characters. 31 characters were eliminated after this first step for one of the following reasons:

1   absence of the character in >20 species.
2   no variation in the data

3   the impossibility of reliably describing or measuring a structure due to difficulties with its examination or to excessive variation within the same individual
... Preliminary analysis with the 14 discrete characters alone could not resolve
relationships clearly and thus continuous characters were included.'

De Luna (1995) 'In principle, a character should be considered if it shows at least
two distinct character states among the groups studied ... [characters were included
if] classification of their variation into character states was readily obvious.'

Liede (1996) 'Succulent shoots are associated with the reduction of leaves to scales
and a squarose habit. However, an analysis by A. Nicholas and D.J. Goyder (unpublished data), in which all three characters relating to succulence are used, resulted
in cladograms in which the succulent taxa identified as sister groups solely because
of their succulence. Therefore, only two of the three obviously linked characters
have been considered here, and an additional analysis using only one of them has
been performed.'

Mészáros et al. (1996) A second data set was derived from a preliminary data set
in order to avoid distortion when using applied weights (Goloboff, 1993) 'In this
data set we coded linearly ordered multistate characters in a binary additive way.'

Monocotyledons: Systematics and Evolution, volumes 1 and 2

Rudall and Cutler (1995) 'Raphides differ from the larger, solitary styloids which
may (in a few taxa e.g. *Dracaena* and *Nolina*) coexist with raphides in the same
plant, and must therefore be treated as a separate (i.e. non-homologous character).'
Linder and Kellogg (1995) 'Characters were coded as binary, except when it was
apparent that this would misinform ... Although the data set distinguishes between
unknown, variable and inapplicable characters, for this analysis they were all treated
as unknown, although this may be misleading in some cases (Nixon and Davis,
1991).'

Uhl et al (1995) 'Various structural features ... are based on simple patterns, but
the resulting characters are complex and homology is often in question.'

Advances in Legume Systematics, Part 7, Phylogeny

Luckow (1995) 'The fruits of three species of *Parkia* are dehiscent: *P. velutina*, *P.
ulei* and *P. platycephala*. These fruits are structurally quite different from the dehiscent fruits found in the *Dichrostachys* group, and homologies are uncertain. They
were coded as missing in this data set; coding them as homologous or uniquely
derived characters did not affect the tree topology of the ingroup.'

Grimes (1995) 'The following three characters [4,5 and 6] all concern branching.
They are scored as individual characters as 5 and 6 represent modifications of 4.'
'Initially most characters were scored as present–absent characters. The problem
with doing this is that many of these 'characters', particularly those of architecture

and branching pattern, may actually be states of characters. If so, treating them as separate and independent characters is misleading as it presumes that one character may be transformed into any other (Pimentel and Riggins, 1987). As a preliminary part of the analyses, when character distribution on the cladogram indicates that the binary characters are better scored as a multistate character, the characters have been rescored as states of one character. Conversely, if the analysis indicated independent derivation of any state of a presumedly multistate character, that character was divided into present–absent characters.'

Crisp and Weston (1995) 'The inability of Hennig86 to handle partial ambiguity in multistate characters forced us to use binary coding throughout' 'Moreover, this result emphasises the value of the cladistic concept of nested character states and thus cladistic analysis as a systematic tool. For example, when character states for antipodal cell morphology were treated as three independent nominal variables – "normal", "giant", "absent" – their taxonomic distribution seemed grossly incongruent with other characters ... However, when we treated this variation as a series of three, nested character states ("normal" ("giant" ("absent"))) – incongruence disappeared.

Appendix 2.3: Detailed breakdown of character coding

Nominal variable coding

The number of nominal variable characters scored is the overall total [e.g. (NOMINAL VARIABLE + 3) is scored as two nominal variable characters], whereas logically related and inapplicable data (missing) characters are scored as sets, each set being scored only once.

Number of matrices composed totally of nominal variable characters	0
Number of matrices composed totally of bistate characters	4
Number of nominal variable characters found in the strictly bistate matrices	5

Ratio coding

Number of matrices which employ (multiple-structure) ratio coding	9
Total number of (multiple-structure) ratio characters	16
Number of matrices which employ single-structure ratio coding	2
Total number of single-structure ratio characters	2

Inapplicable data coding

Number of matrices which use only an inapplicable data (missing) approach	1
Number of matrices which use only an inapplicable data (multistate) approach	15
Number of matrices which use both an inapplicable data (missing) and an inapplicable data (multistate) approach	13
Total number of inapplicable data (multistate) characters	80

Total number of inapplicable data (missing) characters	26
Number of inapplicable data (multistate) characters which are cryptic	24
Number of inapplicable data (missing) characters which are cryptic	5
Number of inapplicable data (multistate) characters which are continuous	19
Number of inapplicable data (missing) characters which are cryptic	0

Positional coding

Number of matrices which use positional data (nominal) coding	2
Number of matrices which use positional data (composite) coding	1
Total number of positional data (nominal) characters	4
Total number of positional data (composite) characters	3

References

Archie, J.W. (1985) Methods for coding variable morphological features for numerical taxonomic analysis, *Systematic Zoology*, **34**, 326–345.

Armbruster, W.S. (1996) Cladistic analysis and revision of *Dalechampia* sections *Rhopalosylis* and *Brevicolumnae*, *Systematic Botany*, **21**, 209–236.

Axelius, B. (1996) The phylogenetic relationships of the physaloid genera (Solanaceae) based on morphological data, *American Journal of Botany*, **83**, 118–124.

Bateman, R.M., DiMichele, W.A. and Willard, D.A. (1992) Experimental cladistic analysis of anatomically preserved arborescent lycopsids from the carboniferous of Euramerica: an essay of paleobotanical phylogenetics, *Annals of the Missouri Botanical Garden*, **79**, 500–559.

Brower, A.V.Z. and Schawaroch, V. (1996) Three steps of homology assessment, *Cladistics*, **12**, 265–275.

Bryant, H.N. (1989) An evaluation of cladistic and character analysis as hypothetico-deductive procedures, and the consequences for character weighting, *Systematic Zoology*, **38**, 214–227.

Chappill, J.A. (1989) Quantitative characters in phylogenetic analysis, *Cladistics*, **5**, 217–234.

Chappill, J.A. and Maslin, B.R. (1995) A phylogenetic assessment of tribe Acacieae, in Crisp, M.D. and Doyle, J.J. (eds) *Advances in Legume Systematics*, Part 7, *Phylogeny*, Kew: Royal Botanic Gardens, pp. 77–100.

Cox, P.A., Huynh, K.-L. and Stone, B.C. (1995) Evolution and systematics of Pandanaceae, in Rudall, P.J., Cribb, P.J., Cutler, D.F. and Humphries, C.J. (eds) *Monocotyledons: Systematics and Evolution*, vol. 1, Kew: Royal Botanic Gardens, pp. 663–684.

Cranston, P.S. and Humphries, C.J. (1988) Cladistics and computers: a chironomid conundrum? *Cladistics*, **4**, 72–92.

Crisp, M.D. and Doyle, J.J. (eds) (1995) *Advances in Legume Systematics*, Part 7, *Phylogeny*, Kew: Royal Botanic Gardens.

Crisp, M.D. and Weston, P.H. (1995) Mirbelieae, in Crisp, M.D. and Doyle, J.J. (eds) *Advances in Legume Systematics*, Part 7, *Phylogeny*, Kew: Royal Botanic Gardens, pp. 245–282.

De Luna, E. (1995) The circumscription and phylogenetic relationships of the Hedwigiaceae (Musci), *Systematic Botany*, **20**, 347–373.

De Pinna, M.C.C. (1991) Concepts and tests of homology in the cladistic paradigm, *Cladistics*, **7**, 367–394.

Farris, J.S. (1990) Phenetics in camouflage, *Cladistics*, **6**, 91–100.

Felsenstein, J. (1988) Phylogenies and quantitative characters, *Annual Review of Ecology and Systematics*, **19**, 445–471.

Goldblatt, P. (1995) The status of R. Dahlgren's orders Liliales and Melanthiales, in Rudall, P.J., Cribb, P.J., Cutler, D.F. and Humphries, C.J. (eds) *Monocotyledons: Systematics and Evolution*, vol. 1, Kew: Royal Botanic Gardens, pp. 181–200.

Goldblatt, P. and Manning, J.C. (1995) Phylogeny of the African genera *Anomatheca* and *Freesia* (Iridaceae: Ixioideae) and a new genus *Xenoscapa*, *Systematic Botany*, **20**, 161–178.

Graham, S.A. (1995) Systematics of *Woodfordia* (Lythraceae), *Systematic Botany*, **20**, 482–502.

Grimes, J. (1995) Generic relationships of Mimosoideae tribe Ingeae, with emphasis on the New World *Pithecellobium* complex, in Crisp, M.D. and Doyle, J.J. (eds) *Advances in Legume Systematics*, Part 7, *Phylogeny*, Kew: Royal Botanic Gardens, pp. 101–122.

Guala, G.F., II (1995) A cladistic analysis and revision of the genus *Apoclada* (Poaceae: Bambusodae), *Systematic Botany*, **20**, 207–223.

Gustafsson, M.H.G. and Bremer, K. (1995) Morphology and phylogenetic relationships of the Asteraceae, Calyceraceae, Campanulaceae, Goodeniaceae and related families (Asterales), *American Journal of Botany*, **82**, 250–265.

Hawkins, J.A. (1996) *Systematics of Parkinsonia L. and Cercidium Tul. (Leguminosae: Caesalpinioideae)*, DPhil thesis, University of Oxford.

Hawkins, J.A., Hughes, C.E. and Scotland, R.W. (1997) Primary homology assessment, characters and character states, *Cladistics*, **13**, 275–283.

Herendeen, P.S. (1995) Phylogenetic relationships of the tribe Swartzieae, in Crisp, M.D. and Doyle, J.J. (eds) *Advances in Legume Systematics*, Part 7, Phylogeny, Kew: Royal Botanic Gardens, pp. 123–132.

Hufford, L. (1996) Developmental morphology of female flowers of *Gyrostemon* and *Tersonia* and floral evolution among Gyrostemonaceae, *American Journal of Botany*, **83**, 1471–1487.

Karis, P.O. (1995) Cladistics of the subtribe Ambrosiinae (Astereaceae: Heliantheae), *Systematic Botany*, **20**, 40–54.

Kellogg, E.A. and Linder, H.P. (1995) Phylogeny of Poales, in Rudall, P.J., Cribb, P.J., Cutler, D.F. and Humphries, C.J. (eds) *Monocotyledons: Systematics and Evolution*, vol. 1, Kew: Royal Botanic Gardens, pp. 511–542.

Kress, W.J. (1995) Phylogeny of the Zingiberanae: morphology and molecules, in Rudall, P.J., Cribb, P.J., Cutler, D.F. and Humphries, C.J. (eds) *Monocotyledons: Systematics and Evolution*, vol. 1, Kew: Royal Botanic Gardens, pp. 443–460.

Lammers, T.G. (1996) Phylogeny, biogeography and systematics of the *Wahlenbergia fernandeziana* complex (Campanulaceae: Campanuloideae), *Systematic Botany*, **21**, 397–415.

Lane, M.A. and Hartman, R.L. (1996) Reclassification of North American *Haplopappus* (Compositae: Astereae) completed: *Rayjacksonia* gen. nov., *American Journal of Botany*, **83**, 356–370.

Lewis, G.P. and Schrire, B.D. (1995) A reappraisal of the *Caesalpinia* group (Caesalpinioideae: Caesalpinieae) using phylogenetic analysis, in Crisp, M.D. and Doyle, J.J. (eds) *Advances in Legume Systematics*, Part 7, *Phylogeny*, Kew: Royal Botanic Gardens, pp. 41–52.

Liede, S. (1996) *Sarcostemma* (Asclepiadaceae) – a controversial generic circumscription reconsidered: morphological evidence, *Systematic Botany*, **21**, 31–44.

Linder, H.P. and Kellogg, E.A. (1995) Phylogenetic patterns in the commelinid clade, in Rudall, P.J., Cribb, P.J., Cutler, D.F. and Humphries, C.J. (eds) *Monocotyledons: Systematics and Evolution*, vol. 1, Kew: Royal Botanic Gardens, pp. 473–496.

Luckow, M. (1995) A phylogenetic analysis of the *Dichrostachys* group (Mimosoideae: Mimoseae), in Crisp, M.D. and Doyle, J.J. (eds) *Advances in Legume Systematics*, Part 7, *Phylogeny*, Kew: Royal Botanic Gardens, pp. 63–76.

Luckow, M. and Hopkins, H.C.F. (1995) A cladistic analysis of *Parkia* (Leguminosae: Mimosoideae), *American Journal of Botany*, **82**, 1300–1320.

Lutzoni, F.M. and Brodo, I.M. (1995) A generic redelimitation of the *Ionaspis-Hymenelia* complex (Lichenized Ascomycotina), *Systematic Botany*, **20**, 224–258.

Maddison, W.P. (1993) Missing data versus missing characters in phylogenetic analysis, *Systematic Biology*, **42**, 576–581.

Mészáros, S., De Laet, J. and Smuts, E. (1996) Phylogeny of temperate Gentianaceae: a morphological approach, *Systematic Botany*, **21**, 153–168.

Nelson, G. (1994) Homology and systematics, in Hall, B.K. (ed.) *Homology: the Hierarchical Basis of Comparative Biology*, San Diego, CA: Academic Press, pp. 101–149.

Patterson, C. (1982) Morphological characters and homology, in Joysey, K.A. and Friday, A.E. (eds) *Problems in Phylogenetic Reconstruction*, London: Academic Press, pp. 21–74.

Patterson, C. and Johnson, G.D. (1997) The data, the matrix and the message: comments on Begle's 'Relationships of the Osmeroid Fishes', *Systematic Biology*, **46**, 458–465.

Pimentel, R.A. and Riggins, R. (1987) The nature of cladistic data, *Cladistics*, **3**, 201–209.

Platnick, N.I. (1979) Philosophy and the transformation of cladistics, *Systematic Zoology*, **28**, 537–546.

Platnick, N.I., Griswold, C.E. and Coddington, J.A. (1991) On missing entries in cladistic analysis, *Cladistics*, **7**, 337–343.

Pleijel, F. (1995) On character coding for phylogeny reconstruction, *Cladistics*, **11**, 309–315.

Pogue, M.G. and Mickevich, M.F. (1990) Character definitions and character state delineation: the *bête-noire* of phylogenetic inference, *Cladistics*, **6**, 319–361.

Rieppel, O.C. (1988) *Fundamentals of Comparative Biology*, Basel: Birkhäuser Verlag.

Ronse-Decraene, L.P., De Laet, J. and Smets, E.F. (1996) Morphological studies in Zygophyllaceae. II. The floral development and vascular anatomy of *Peganum harmala*, *American Journal of Botany*, **83**, 201–215.

Rudall, P.J., Cribb, P.J., Cutler, D.F., and Humphries, C.J. (1995) *Monocotyledons: Systematics and Evolution*, volumes 1 and 2, Kew: Royal Botanic Gardens.

Rudall, P.J. and Cutler, D.F. (1995) Asparagales: a reappraisal, in Rudall, P.J., Cribb, P.J., Cutler, D.F. and Humphries, C.J. (eds) *Monocotyledons: Systematics and Evolution*, volume 1, Kew: Royal Botanic Gardens, pp. 157–168.

Schrire, B.D. (1995) Evolution of the tribe Indigofereae (Leguminosae: Papilionoideae), in Crisp, M.D. and Doyle, J.J. (eds) *Advances in Legume Systematics*, Part 7, *Phylogeny*, Kew: Royal Botanic Gardens, pp. 161–244.

Siebert, D.J. (1992) Tree statistics; trees and 'confidence'; consensus trees; alternatives to parsimony; character weighting; character conflict and its resolution, in Forey, P.L., Humphries, C.J., Kitching, I.J., Scotland, R.W., Siebert, D.J. and Williams, D.M. (eds) *Cladistics: a Practical Course in Systematics*, Oxford: Clarendon Press, pp. 72–88.

Simpson, D. (1995) Relationships within Cyperales, in Rudall, P.J., Cribb, P.J., Cutler, D.F. and Humphries, C.J. (eds) *Monocotyledons: Systematics and Evolution*, vol. 1, Kew: Royal Botanic Gardens, pp. 497–510.

Smith, A.B. (1994) *Systematics and the Fossil Record: Documenting Evolutionary Patterns*, Oxford: Blackwell Scientific Publications.

Smith, J.F. and Doyle, J.J. (1995) A cladistic analysis of chloroplast DNA restriction site variation and morphology for the genera of the Juglandaceae, *American Journal of Botany*, **82**, 1163–1172.

Stevens, P.F. (1991) Character states, morphological variation, and phylogenetic analysis: a review, *Systematic Botany*, **16**, 553–583.

Stevenson, D.W. and Loconte, H. (1995) Cladistic analysis of monocot families, in Rudall, P.J., Cribb, P.J., Cutler, D.F. and Humphries, C.J. (eds) *Monocotyledons: Systematics and Evolution*, vol. 1, Kew: Royal Botanic Gardens, pp. 543–578.

Strait, D.S., Moniz, M.A. and Strait, P.T. (1996) Finite mixture coding: a new approach to coding continuous characters, *Systematic Biology*, **45**, 67–78.

Thiele, K. (1993) The holy grail of the perfect character: the cladistic treatment of morphometric data, *Cladistics*, **9**, 275–304.

Uhl, N.W., Dransfield, J., Davis, J.I., Luckow, M.A., Hansen, K.S., and Doyle, J.J. (1995) Phylogenetic relationships among palms: cladistic analyses of morphological and chloroplast DNA restriction site variation, in Rudall, P.J., Cribb, P.J., Cutler, D.F. and Humphries, C.J. (eds) *Monocotyledons: Systematics and Evolution*, vol. 1, Kew: Royal Botanic Gardens, pp. 623–662.

Weller, S.G., Wagner, W.L. and Sakai, A.K. (1995) A phylogenetic analysis of *Schiedea* and *Alsinidendron* (Caryophyllaceae: Alsinoidea): implications for the evolution of breeding systems, *Systematic Botany*, **20**, 315–337.

Wilkinson, M. (1995) A comparison of two methods of character construction, *Cladistics*, **11**, 297–308.

Van Wyk, B.-E. and Schutte, A.L. (1995) Phylogenetic relationships in the tribes Podalyrieae, Liparieae and Crotalarieae, in Crisp, M.D. and Doyle, J.J. (eds) *Advances in Legume Systematics*, Part 7, *Phylogeny*, Kew: Royal Botanic Gardens, London, pp. 41–52.

Chapter 3

Experiments in coding multistate characters

Peter L. Forey and Ian J. Kitching

Introduction

Understanding biological diversity by means of classifications requires that we translate our observations into characters. Characters are the units of language through which we communicate our ideas of homology, relationship, diagnosis and identity. Although most biologists would agree with this as a general statement, it is equally true that different workers would partition their observations into different units or characters (Smith, 1994). Sometimes these differences are purely practical, commensurate with a particular method of systematic analysis. For example, a pheneticist would accept a ratio (e.g. 0.5) or a simple length measurement (e.g. 6.2 mm) as a character. In contrast, a cladist, because of the demands of the method, needs to translate observations to discrete variables (Archie, 1985; Thiele, 1993). Different workers may also choose to describe shape, or mutual relationships of bones or insect wing venation, in different ways. These relatively trivial differences are usually transparent. But beneath this veneer of variation there is a more fundamental core of differences as to what we mean by a character and this often implicitly determines how we code our observations in any cladistic analysis. Furthermore, because the coding method that we use is the prime determinant of the systematic outcome, particular attention to this aspect of cladistic analysis is necessary (Pimentel and Riggins, 1987; Bryant, 1989; Pogue and Mickevich, 1990; Hawkins *et al.*, 1997).

Definitions

Within the context of cladistic analysis, there are several definitions of the concept of a character. Those definitions fall into three main groups, exemplified by the following quotes.

1 'Any attribute of an organism or a group of organisms by which it differs from an organism belonging to a different category or resembles an organism of the same category' (Mayr *et al.*, 1953).
2 'A character is a feature of an organism which is the product of an ontogenetic or cytogenetic sequence of previously existing features or a feature of a previously existing parental organism(s). Such features arise in evolution by the modification of previously existing ontogenetic or cytogenetic or molecular sequences' (Wiley, 1981).

3   'Cladistics is a discovery procedure, and its discoveries are characters (homologies) and taxa' (Nelson and Patterson, 1993).

The first definition is operational in outlook and views characters as static observations that are useful both to diagnose taxa from one another and to group taxa together on the basis of structural identity. In its purest form, such a definition does not distinguish between characters and character states.

The second definition is the transformational view of a character. This is much more subjective and views characters as transformation series of character states. Character states are the observations, while the character is a particular series of transformations intuitively linked. That is, structures that do not look the same or function the same are nevertheless considered to be part of a single transformation series because they are linked by process through time. In the particular definition cited here (Wiley, 1981) that process is ontogeny, which may be observable. But other transformational definitions do not specify ontogeny, only that structures existing earlier in time (phylogenetically or ontogenetically) transform into other structures; that is, character states have an ancestral–descendent relationship to one another within a given transformation.

The third definition of a character is what may be called the modern cladist's view, namely that characters are constructed as hypotheses and are subject to the cladistic test of congruence in a parsimony analysis. This is the definition of a character that many, if not most, would subscribe to. It is not incompatible with the other two but it does view characters from a different perspective. Three item analysis is an extension of this view of a character (see Chapters 8 and 9). Certainly it differs from the transformational view in that it excludes process from the definition.

From this brief list of definitions, it is clear that 'character' is a much deeper concept than some notion of simple or direct observation. These different ideas lead to different methods of coding, which in turn can lead to different systematic outcomes. Several authors have pointed out that there is a wealth of literature about how we analyse characters but a dearth on how we code those characters in the first place. When looking through cladistic literature, it is clear that there is considerable variation in the way that authors code their observations (see Chapter 2). In reality the most intractable barrier to settling disputes about systematic conclusions of a particular group of organisms usually centres on the coding of characters rather than methods of analysis (e.g. Gauthier et al., 1988 vs Gardiner, 1982, 1993).

Coding practices

For the first and third type of character definitions given above, the usual way to code observations is as simple presence/absence (+/–): a structure is either present or absent, or it is blue or it is red, i.e. binary coding. Any theory of transformation is an a posteriori exercise deduced from the optimization of the characters onto the selected cladogram. Binary coding is also compatible with the second definition in which implied transformation is one of ontogenetic/phylogenetic gain or loss. There is usually little dispute about the coding of such characters.

Problems arise, however, when the investigator believes, perhaps with good reason, that certain attributes are linked, logically and/or biologically (Wilkinson, 1995), and that it is desirable to express such linkage as a multistate character. Perhaps the most obvious multistate character is an identified nucleotide position within a DNA/RNA molecule, where the states are ACG(U)T(+absence), yielding a 4(5)-state character. There is little dispute among molecular systematists as to how we should code nucleotide characters (although there is much discussion about how those data are then analysed). However, here we concern ourselves specifically with the coding of morphological data, which, we believe, is a much more complex issue. Morphological multistate characters may be coded for in several different ways (Pimentel and Riggins, 1987; Pleijel, 1995; see Chapter 2). These procedures contain certain assumptions about the relations between the character states (i.e. about the evolution of that character) that can lead to different consequences for the systematic outcome. The remainder of this chapter is concerned with a consideration of the multistate character.

Coding methods

Pleijel (1995) listed four types of coding, pertinent to a hypothetical example of shape and colour of a particular structure and the presence/absence of that structure. His example is reproduced here as Table 3.1. Pleijel suggested that there are four ways in which these variables may be described for analysis. Method A is the pure multistate, in which every combination of presence/absence/black/white/square/round is given a unique numeric value. All the observations relating to the feature and the attributes colour and shape are included in a single row of data. In method B, the attributes of colour and shape are separated as two separate multistate characters with the absence of each scored as '0'. In method C, each of the attributes colour and shape is treated as a separate character (i.e. as a separate row of data), while there is a third character that codes for the presence/absence of the structure as a whole. Those taxa that do not possess the coloured or shaped feature are coded as non-applicable for the two characters relating to shape and colour. Computationally this non-applicable coding is accommodated by means of a question mark. Method C is one most often used because it is intuitively obvious. It has been called 'conventional coding' by Hawkins et al. (1997; see Chapter 2). We call it contingent coding here because the coding of one character is contingent upon that of another (i.e., whether we can code for characters 2 and 3 depends upon the coding in character 1). Method D2 treats every observation as a single character in which the variable is either there or it is not. In this example this results in five rows of data: row 1 specifies whether the feature is there or not; row 2 indicates whether it is square or not; row 3 indicates whether the circle shape is present or not, etc. This is often called presence/absence coding (Pleijel, 1995) or nominal variable coding (Pimentel and Riggins, 1987), but Wilkinson (1995) referred to it as reductive coding because it reduces the observed variation to simple characters that each indicate whether a particular feature is there or not.

Pleijel admitted that he had not exhausted the possibilities of coding and there are further methods of coding these data (D1, E and F in Table 3.1). Here method

Table 3.1 Coding methods (based on Pleijel, 1995)

	–	□	○	■	●
Method A					
multistate	0	1	2	3	4
Method B					
two multistates					
character 1 shape	0	1	2	1	2
character 2 colour	0	1	1	2	2
Method C					
contingent coding					
character 1 presence/absence of feature	0	1	1	1	1
character 2 shape	?	0	1	0	1
character 3 colour	?	0	0	1	1
Method D					
D1 presence/absence of particular combination					
character 1 presence/absence of feature	0	1	1	1	1
character 2 square–white	0	1	0	0	0
character 3 square–black	0	0	0	1	0
character 4 round–white	0	0	1	0	0
character 5 round–black	0	0	0	0	1
D2 presence/absence of individual attribute					
character 1 presence/absence of feature	0	1	1	1	1
character 2 square	0	1	0	1	0
character 3 round	0	0	1	0	1
character 4 white	0	1	1	0	0
character 5 black	0	0	0	1	1
Method E					
character 2 square	0	1	0	1	0
character 3 round	0	0	1	0	1
character 4 white	0	1	1	0	0
character 5 black	0	0	0	1	1
Method F					
character 2 square	?	1	0	1	0
character 3 round	?	0	1	0	1
character 4 white	?	1	1	0	0
character 5 black	?	0	0	1	1

E deletes the first column of data from the presence/absence coding because the attributes of shape (round or square) or colour (black or white) cannot be found in isolation. They are descriptors of the feature 'shape' and the row representing the presence of the feature may be considered to be redundant. Another method (F) uses a combination of presence/absence and non-applicable (assigned a question mark) for those taxa that lack the feature being described. This is an attempt to distinguish between the impossibility of having black/white colour or round/square shape, because the structure is not there, from the observation that the structure is there but a particular descriptor is not.

Implications in a real example

These different methods of coding may lead to different phylogenetic conclusions. Before returning to the above abstract example (Table 3.1) we present a real example, to show the consequences of two of these different coding methods. For this example we have used a data matrix published by Wake (1993), to establish relationships amongst caecilians (gymnophionians). The characters were derived from observations of the neuroanatomical system and concerned the eye, ear, hypoglossal nerve, olfactory and vomero-nasal organs. We chose this data set for several reasons. First, some of the characters used by Wake (1993, 1994) are complex, with many included states, and therefore satisfy our desire to explore the multistate character and alternative codings. Second, Wake (1993: 44) stated unambiguously why she linked many observations into a multistate character: 'I coded character states in a moderately unconventional way, but one logical given the data: in such cases, 0 is the presence (or absence) of a state, and modifications of the alternative state that are not clearly innovations, but may be developmental modifications (often heterochronies), are coded as sequences of the alternative state (e.g. 1a, 1b, etc.)'. Thus, Wake's view of a character agreed with the second class of definitions outlined above. Third, Wake was particularly keen to introduce new (non-traditional) morphological data to help refine a systematic problem in need of more characters. Fourth, Wake (1994: 187) recognized the problem of character dependence/independence. Fifth, Wake (1993: 43) clearly pointed out that analysis of these characters was preliminary, thereby inviting comment (which has already been forthcoming (Wilkinson, 1995, 1997)). It is not our intention to criticize the observations or systematic conclusions given by Wake. Wilkinson (1995, 1997) discussed alternative codings for Wake's data. In part, Wilkinson's alternatives (1995) relate to the methods by which the initial observations were coded, something that we are also concerned with here. He also clarified some of the ambiguous codings originally given by Wake (1993: Appendix 1); for instance, the coding of the attachment pattern of a lens (eye character 13, code 1b) in organisms (*Afrocaecilia* and *Boulengerula*) that have no lens (eye character 12, code 1f). We are not concerned with such discussion here since we are not competent to comment on the original observations. Rather we use some of these characters to illustrate several aspects of character coding that have generally been dealt with in abstract terms (Hawkins *et al.*, 1997; Pimentel and Riggins, 1987; Pleijel, 1995).

Wake's (1993) original data set contained 26 taxa and she divided her characters into four data sets (eye, ear, hypoglossal nerve and olfactory/vomeronasal organs). However, not all taxa were represented in all four data sets and in some cases there were question marks against some of the cell entries. Reducing the taxon sampling to those taxa that are completely known and represented in all four data sets results in a new, reduced data set of 10 taxa (Appendices 3.1 and 3.2). Of the original 34 characters, 15 are multistate, of which some in the original matrix (Wake, 1993) have seven, eight or even nine states. However, because of our reduced taxon sampling here, a maximum of five states is represented (Appendix 3.1, character 1). The reduced taxon data set for all 34 characters is represented in Appendix 3.1, where the original codes used by Wake (0, 1a, 1b, etc.) have been translated to 0, 1, 2, etc.

A recoded data set for the same taxa has thus been created, using the extreme version of presence/absence coding (Appendix 3.2). In some cases we have recoded according to the presence/absence of constituent parts of the character. For instance, character number 1 in the original data set specifies particular combinations of the presence/absence of six eye muscles as well as the shape of one of those muscles. Wilkinson (1995) has already discussed this character. We treat the presence/absence of a particular eye muscle as a single character and therefore this original multistate becomes seven individual characters. Similarly, we have recoded the original character 24 (the condition of the hypoglossal nerve) as five separate characters (this character is discussed further below).

For other multistate characters, we have simply recoded a 'particular combination' as an individual character with the states present/absent. For example, character 12 in the original data set relates to the composition of the lens and has seven states, four of which are represented in the taxa included here (Appendix 1A). These are lens crystalline (0), lens cellular peripherally and crystalline medially (1), lens cellular (2), lens amorphous (3). Other combinations (Wake, 1993: 45) not represented in the taxa used here include crystalline peripherally and cellular medially, rudimentary, lens absent. This may seem unsatisfactory but without knowing more about the actual observations it is not evident how one would code for individual constituents (cellularity, crystallinity, etc.) (see Wilkinson (1997) for more extensive discussion of this and other similar characters used in the original data set).

Granted that this is an imperfect recoding of the data we recognize 77 characters in the presence/absence matrix (Appendix 3.2). There are few characters in which an entire structure was absent and therefore we do not consider the other methods (B and C above) at this stage. We are only concerned here with the possible effects of different codings on the systematic outcome.

Analysis using Hennig86 (Farris, 1988) with all multistates treated as unordered resulted in eight cladograms, the strict consensus tree of which is shown in Fig. 3.1A. Analysis of the recoded data using presence/absence resulted in four cladograms, the strict consensus of which is shown in Fig. 3.1B. There are only two nodes shared by these two consensus trees (J–E and I + E). However, one of the eight original cladograms for the multistate data set is identical to one of the four cladograms of the presence/absence data set (Fig. 3.1C). The increased resolution to be gained by recoding the data as presence/absence is clearly an advantage, but probably not decisive in that the multistate coding set does contain the phylogenetic signal revealed by the presence/absence data coding. Parenthetically, we would note that one of the eight cladograms using multistate characters recovers the mutual

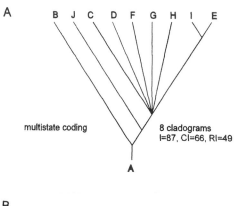

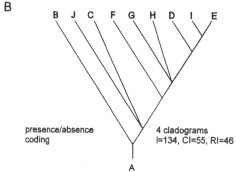

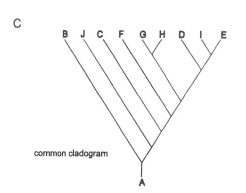

Figure 3.1 A. Strict consensus tree resulting from the analysis of data set given in Appendix 3.1 which includes 15 multistate characters (asterisks). B. Strict consensus tree resulting from the analysis of data set given in Appendix 3.2, which is the data given in 1A but recoded as presence/absence coding. C. There is one cladogram common to the two analyses.

relationships of those taxa in Wake's analysis of all 26 taxa (Wake, 1993: Fig. 5), so this may mean that it is a representative subset of taxa.

The differences between the methods of coding are more evident when the data set is partitioned into eye characters, ear characters, hypoglossal nerve characters and olfactory/vomeronasal characters (Fig. 3.2). We will detail just the effects

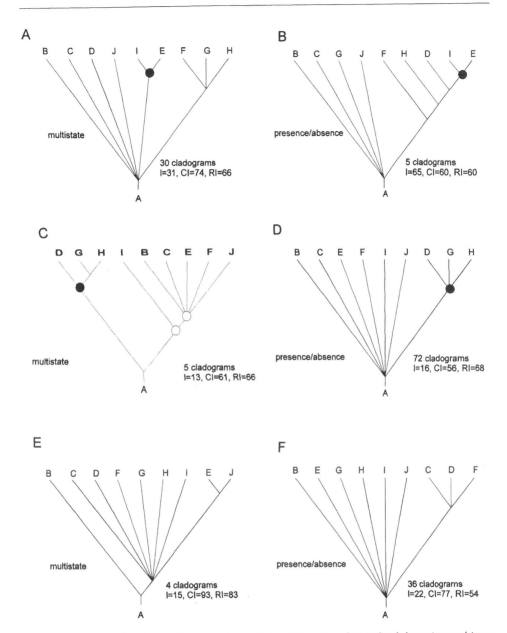

Figure 3.2 Results of analysing three portions of the data set in Appendix I by using multistate coding and translation of the data into presence/absence coding. In each pair of strict consensus trees the multistate is on the left and the presence/absence coding to the right. A, B, eye characters. C, D, hypoglossal nerve characters. E, F, ear characters. For completeness we record that olfactory/vomero-nasal characters gave completely unresolved consensus trees using both techniques and are not shown here.

on the eye and hypoglossal nerve characters since these contain characters (Wake's 1993 original eye character 1 and hypoglossal nerve character 1) that could be recoded without any further knowledge of the animals. We have also included the effects on the ear characters but ignored the vomero-nasal characters, because analysis of those data leads to complete non-resolution in both types of coding.

The effect of the two different methods of coding for the eye characters is shown in Figs 3.2A and B, the hypoglossal nerve characters in Figs 3.2C and D, and the ear characters in Figs 3.2E and F. For the eye characters the presence/absence coding gave fewer original cladograms and greater resolution of the consensus (Fig. 3.2B), while for the hypoglossal nerve characters and the ear characters the opposite was true (Figs 3.2C, E). Thus, there is no direct law-like behaviour of number of characters leading to greater resolution or fewer cladograms. More importantly, different groups are specified by differently-coded data sets. For instance, in the eye data sets, the multistate coding recognises a group F+G+H whereas this clade does not appear either in the consensus or among the fundamental cladograms of the presence/absence data set. Thus, these methods of coding lead to completely contradictory results.

For the hypoglossal nerve characters one node (D+G+H) is common to the two consensus trees but here the nodes denoted by the open circles do appear in some of the 72 fundamental cladograms from which the consensus is derived. For the ear characters, then, one of the 36 fundamental cladograms from analysis of the presence/absence coding is represented by one of the four fundamental cladograms of the multistate data. The problem is that we have no reason to choose these particular cladograms from among the many. And in any analysis, we choose one or another method of coding, although this may vary from one character to another within a single data matrix (see Chapter 2).

Consequences

Given that there are different ways of coding and that these may lead to different phylogenetic conclusions, we need to be aware of their consequences in at least three areas: character dependence/independence, the nature of implied transformations (optimization), which is a goal of standard cladistic analysis (cf. three-item statements analysis; see Chapters 8 and 9), and resolution (which we have partly covered above).

Pleijel (1995) provides our backbone for discussion (Table 3.1). In method A, the complete multistate character, the states are treated as linked into a single transformation series. Different states represent different contributions of attributes, of which one of the states is absence. This is comparable with some of the characters used by Wake (1993, 1994). As Pleijel (1995) noted, the advantage of such a multistate character is that it avoids the possibility that false homologies will obscure a phylogenetic signal justified by true homologies. Since each row of data is assumed to be an independent estimator of phylogeny and, in the absence of character weighting, is also assumed to contribute equally to parsimony analysis, then the multistate character avoids the problem of logical dependency (see below). However, linking together different observations denies the possibility of testing propositions of homology between the states because the test of homology involves comparing one row of data with another (Patterson, 1982; de Pinna, 1991).

The multistate character also suffers from problems of transformation. In this theoretical example, we can treat the character as ordered or unordered. Both lead to unsatisfactory theories of character transformation. For instance, treating the character as an ordered multistate implies that we 'know' the order (e.g. that a white square always precedes or follows a white circle). In limited instances there may be ontogenetic evidence of both order and polarity and if this were the case certainly the ordered multistate would be advantageous. But such circumstances are extremely rare. Another consequence of the ordered multistate is that an implied transformation $\square \rightarrow \blacksquare$ costs (in terms of numbers of steps) the same as $- \rightarrow \bigcirc$. In other words, a single change in colour is equal to the acquisition of both a shape and a colour.

Treating the multistate as unordered would treat the transformations $\blacksquare \rightarrow \bigcirc$ and $\blacksquare \rightarrow \square$ as the same cost. Yet for the first transformation both shape and colour have changed, whereas for the second only colour has changed.

Coding method B, where shape and colour are treated as separate rows of data, potentially allows shape and colour to vary independently. But once again, regarding these as ordered multistates implies that we 'know' the transformation and it also implies that $- \rightarrow \blacksquare$ is equal to three steps whereas to transform $- \rightarrow \square$ costs two steps. There is also redundancy in this type of coding in the sense that the acquisition of the structure appears twice – once gaining a colour and once gaining a shape. But, more logically, we suggest that if the structure is gained, it will have both a colour and a shape.

In coding method C – the contingent coding method – the variation is split into three characters. The first denotes the presence/absence of the feature, the second denotes shape and the third denotes colour. Those taxa that do not possess the feature are coded as not applicable for colour and shape and for computational purposes are assigned a question mark. This is the method favoured by Hawkins *et al.* (1997), who argue that it is to be preferred because the observations of shape (square or circle) and the observations of colour (white or black) must be regarded as variables of the same thing (the feature) and are not independent. In fact, it may be argued that we do not need the first character under this coding because the non-applicable coding in characters 2 and 3 implies absence. The drawback to this method resides in the question marks, which lead to spurious optimizations, as well as to over-resolved and spurious maximally parsimonious cladograms (Maddison, 1994).

Coding method D, presence/absence coding (Pleijel, 1995) or reductive coding (Wilkinson, 1995), treats all identical observations as separate characters. In other words, the feature is either there or not, it is white or not, it is square or not, etc. In the resulting matrix there are five rows of data: the first denotes whether the structure is there or not and the others record the presence/absence of the individual attributes. This type of coding breaks the linkage between the attributes of colour and shape. In this particular example, either colour or shape is homoplastic with respect to the other attribute. This homoplasy would be disguised in the multistate coding. In other words, we are not recognizing phylogenetic signals provided by all our observations.

A real example is provided by the first hypoglossal nerve character in the Wake data set (Table 3.2). This refers to the composition of the hypoglossal nerve which

Table 3.2 Taxonomic distribution of elements which make up the hypoglossal nerve. In the original matrix (Wake, 1993, hypoglossal nerve character 1) the character was coded as a multistate character with eight states, as shown in the last column. The letter abbreviations given in the first column are those used in Appendices 3.1 and 3.2. If the columns are treated as separate characters as in presence/absence coding, homoplasy is revealed which is otherwise hidden in the multistate coding. Thus: spinal 2 is incongruent with spinal 1; spinal 3 is incongruent with spinal 1; spinal 3 is consistent with spinal 2; occipital is consistent with spinal 1; occipital is incongruent with spinal 2; vagus is incongruent with spinal 1; vagus is consistent with spinal 2; vagus is consistent with occipital

Taxon	Composition of hypoglossal nerve					
	Spinal 1	Spinal2	Spinal3	Occipital	Vagus	Multistate code
K Epicrionops bicolor	+	+				0
A Epicrionops petersi l	+	+	+			1a
L Ichthyopohis sp. (larva)	+	+				0
M Ichthyophis kohtaoensis	+	+	+	+		1d
N Uraeotyphlus narayani	+	+	+	+		1d
B Dermophis mexicanus	+	+		+		1b
C Gymnopis multiplicata	+	+		+		1b
D Caecilia occidentalis	+	+			+	1f
E Oscaecilia ochrocephala	+	+				0
O Siphonops annulatus	+	+				0
P Boulengerula boulengeri		+			+	1g
F Geotryptes seraphini	+	+		+		1b
H Idiocranium russelli	+	+			+	1f
G Hypogeophis rostratus	+	+				0
Q Grandisonia alternans	+	+				0
R Gegeneophis ramaswamii		+	+			1h
I Scopelomorphus ulugurensis	+	+		+		1b
S Scopelomorphus vittatus	+	+		+		1b
J Typhlonectes natans	+	+		+		1b
T Chthonerpeton indistinctum	+	+				0
U Nectocaecilia haydeii	+			+		1c

supplies the tongue (see also Wilkinson, 1995). The hypoglossal nerve is a complex nerve that is composed of several components (the first three spinal nerves, the occipital and a branch of the vagus). In different species of caecilians, different numbers and different combinations of these nerve components make up the hypoglossal nerve. As an analogy, this is rather like telephone cables that are outwardly similar but with some cables containing red, blue, and black wires, others containing blue yellow and green wires and yet others contain only blue and red wires.

Wake decided to code the composition of the hypoglossal nerve as a complex multistate character with eight states, each state being recognized by a particular combination of wires. Wilkinson (1995) discussed problems associated with ordering or unordering such a multistate character. Here we wish to point out other problems. Translating the multistate coding into presence/absence coding demonstrates clearly that there is homoplasy between the observations of inclusion and exclusion of three constituent wires – the inclusion of spinal 3 in the hypoglossal nerve is homoplastic with respect to the inclusion/exclusion of spinal 1 and the occipital nerve. The occurrence of these three wires specifies different groups (see also the legend to Table 3.2). Such information is lost in multistate coding. Also, in this example the presence/absence coding reveals the interesting fact that the inclusion of the occipital or vagus nerves specifies mutually exclusive groups (see also the legend to Table 3.2). This raises the possibility that, through this method of coding, we may have discovered a new transformation concealed in the multistate coding, which is that spinal 3 has replaced the vagus or vice versa as a constituent of the hypoglossal nerve.

Presence/absence coding is more in keeping with the character as a theory definition (see the list given under 'Definitions' above). In treating every state as a character, it avoids incorrectly grouping states into one transformation series. It also produces several columns of data (more characters), and this should be beneficial for increasing the severity of the test for character congruence and, hence, homology. However, there are three disadvantages. The first is that there can be redundancy of information. In particular, if there were taxa that lacked the feature altogether, all codings would be zeros for each of the columns that specified presence/absence of that particular attribute. This would tend to force these taxa into basal positions on the cladogram or else lead to a type of 'long branch attraction', in which two taxa are grouped together by large numbers of reversals/losses. The second disadvantage is that the codings of adjacent characters are not independent from one another. For instance, in those taxa that have the feature, the attributes of black or white are complementary and must be so, since the coding of black (character 2) automatically determines the coding of white (character 3). The third unfortunate characteristic centres on the optimizations of character states, which we deal with below.

Speaking in tongues

Following our aim to use real characters wherever possible, we now describe the consequences of using different coding methods on structures associated with the pupal tongue case, which one of us (IJK) had to address in a wider study of the phylogeny of the moth family Sphingidae.

Moths of the family Sphingidae are commonly known as hawkmoths, a name that is probably derived from the adults' habit of hovering in front of flowers and drinking nectar through their long tongues (or proboscides). During the pupal stage, the developing adult tongue is housed in a case that runs medially along the ventral surface of the body between the legs and wings. In contrast to a number of other lepidopteran families, the long tongue of an adult hawkmoth is not accommodated in the pupa by lengthening the pupal tongue case beyond the end of the forewings. In the subfamily Sphinginae, the additional length is contained in a separate, free tongue case (FTC), which projects from the antero-ventral surface of the pupa (Figs 3.3B–F). Within the FTC, the adult tongue is turned back on itself once and the apex of the FTC is slightly bulbous in order for this to be accomplished without imposing too tight a bend on the developing proboscis. The extra length gained is thus equivalent to twice the length of the FTC.

Four basic types of FTC can be recognized. Genera such as *Sphinx* have a short FTC that is closely adpressed to the ventral surface of the pupa (Fig. 3.3B). In contrast, the adpressed FTC of *Panogena* is extremely long, reaching the end of the

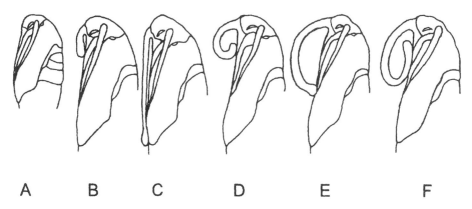

Figure 3.3 Observed conditions of the sphingid pupal free tongue case (FTC). Codings are given in Table 3.3. A, absent (*Clanis, Ceratomia*); B, short, adpressed and non-recurved (*Sphinx, Paratrea*); C, long, adpressed and non-recurved (*Panogena* 1, *Panogena* 2); D, short, looped and non-recurved (*Neogene, Manduca florestan*); E, long, looped and non-recurved (*Manduca sexta, Cocytius*); F, long, looped and recurved (*Agrius* 1, *Agrius* 2).

forewings (Fig. 3.3C). However, in most sphingines, the FTC is looped away from the body of the pupa to produce the characteristic 'jug-handle'. This type can be very short (Fig. 3.3D), as in *Neogene* and some species of *Manduca* (e.g. *M. florestan* and *M. vestalis*), or very long (Fig. 3.3E), as in other species of *Manduca* (e.g. *M. sexta* and *M. hannibal*) and genera such as *Cocytius*. In all of the above, the apex of the FTC is not recurved, being directed either posteriorly (Figs 3.3A and B) or towards the ventral surface of the pupa (Figs 3.3C and D). A few genera have extremely long adult tongues (e.g. *Agrius*). The extra length is accommodated by further lengthening the FTC, then recurving it, so that the apex is directed anteriorly (Fig. 3.3F).

Five methods, A–E (below) were selected to encode the observed variation in the FTC. Methods A–D relate to the protocol outlined in Table 3.1.

Method A: FTC features treated as a single multistate character:

1   absent (0); short, adpressed and non-recurved (1); long, adpressed and non-recurved (2); short, looped and non-recurved (3); long, looped and non-recurved (4); long, looped and recurved (5).

Method B: FTC features of length, shape and curvature treated as three separate multistate characters:

2   absent (0); short (1); long (2)
3   absent (0); adpressed (1); looped (2)
4   absent (0); non-recurved (1); recurved (2).

Method C: contingent coding of FTC features of length, shape and curvature as three independent binary characters and an additional binary character to indicate the presence or absence of a free tongue case; states of characters 6–8 are non-applicable to taxa coded 0 for character 5 and are accommodated using question marks:

5 absent (0); present (1)
6 short (0); long (1)
7 adpressed (0); looped (1)
8 non-recurved (0); recurved (1).

Method D: absence/presence (A/P) coding of FTC features:

 9 free tongue case absent (0); features present (1)
10 short free tongue case absent (0); present (1)
11 long free tongue case absent (0); present (1)
12 adpressed free tongue case absent (0); present (1)
13 looped free tongue case absent (0); present (1)
14 non-recurved free tongue case absent (0); present (1)
15 recurved free tongue case absent (0); present (1).

Method E: Sankoff coding

Sankoff coding: FTC features treated as a single multistate character with transformations between states coded using a Sankoff matrix:

16 absent (0); short, adpressed and non-recurved (1); long, adpressed and non-recurved (2); short, looped and non-recurved (3); long, looped and non-recurved (4); long, looped and recurved (5).

We also included five additional binary characters (17–21), which we refer to as the topological constraint characters (TCCs). This was done because we are trying to explore the effect of multistate characters alongside binary characters, which is a situation usually met with in any real analysis. Analysed separately, these topological constraint characters produce the topology shown in Fig. 3.4. The TCCs were included to assess the influence of additional characters on the results obtained from the various coding methods used for the FTC.

Each data set was analysed, with and without the TCCs, using the exact options of Hennig86 version 1.5 (ie*; Farris, 1988); PAUP version 3.1 (branch and bound; Swofford, 1993) and NONA version 1.50 (amb-; wh; mswap+; Goloboff, 1996). We record the results against each of the coding methods.

Method A: a single multistate character (character 1)

There is no justification for regarding the states of character 1 as forming an ordered sequence, and so this character was treated as unordered. When character 1 was analysed alone, Hennig86 found 5811 equally most-parsimonious cladograms (MPCs) (the memory limit for the PC used), NONA found 1000 (the program limit), while the PAUP analysis was aborted after more than 2000 cladograms had been found. Each of these MPCs is five steps long, with an ensemble consistency

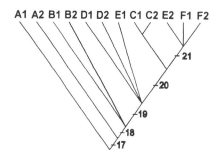

Figure 3.4 The most-parsimonious tree produced from analysis of the five binary topological constraint characters (TCCs), 17–21, rooted using taxon A1 (*Clanis*) as the outgroup.

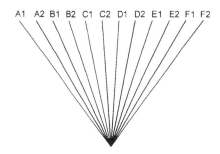

Figure 3.5 The strict consensus tree (an uninformative bush) derived from the numerous equally most-parsimonious trees found from analysis of the single unordered, multistate character (method A).

index (CI) of 1.00 and an ensemble retention index (RI) of 1.00. The strict consensus tree of the results from each analysis is an uninformative bush (Fig. 3.5).

When the TCCs were included with character 1, Hennig86 found six MPCs (length = 10; CI = 1.00; RI = 1.00). Three of these (Fig. 3.6A–C) were also found by NONA and are strictly-supported cladograms (Nixon and Carpenter, 1996). In the other three MPCs, an alternative (DELTRAN; Swofford, 1993) optimization of the 'origin' of character state 1(1) results in taxa B1 and B2 forming a group. PAUP found 95 MPCs of 10 steps, which were not investigated further. The strict consensus tree of the three strictly-supported MPCs (Fig. 3.6D) has a length of 11 steps. The extra step is due to there being two alternative, non-combinable optimizations of character states 1(3) and 1(4). In Fig. 3.6A, character state 1(3) supports the group CDEF and state 1(4) represents a subsequent transformation that separates the group CEF from taxa D1 and D2, which form a paraphyletic pair. In contrast, in Fig. 3.6C, it is character state 1(4) that supports the group CDEF, and 1(3) that is the subsequent transformation, this time supporting the monophyly of D1+D2. The cladogram in Fig. 3.6B is a kind of 'hybrid' between Figs 3.6A and C, in which character state 1(4) can support either the group CDEF (position indicated in square brackets) or the group CEF. The cladograms in Figs 3.6A and B, or those in Figs 3.6B and C, can be combined into a

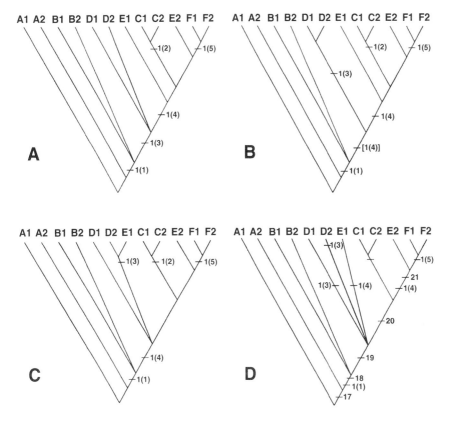

A1 A2 B1 B2 D1 D2 E1 C1 C2 E2 F1 F2

A

1(2)
1(5)
1(4)
1(3)
1(1)

A1 A2 B1 B2 D1 D2 E1 C1 C2 E2 F1 F2

B

1(2)
1(5)
1(3)
1(4)
[1(4)]
1(1)

A1 A2 B1 B2 D1 D2 E1 C1 C2 E2 F1 F2

C

1(3) 1(2) 1(5)
1(4)
1(1)

A1 A2 B1 B2 D1 D2 E1 C1 C2 E2 F1 F2

D

1(3)
1(5)
1(3) 1(4)
21
1(4)
20
19
18
1(1)
17

Figure 3.6 A–C The three equally most-parsimonious trees found by analysis of the single, unordered, multistate character I and the TCCs using NONA. However, only those in (A) and (C) are strictly supported trees (see text for details). D. The strict consensus tree of the three trees in A–C.

strict consensus tree without any increase in length. In other words, there are really only two strictly-supported cladograms: Figs 3.6A and C.

In an unordered multistate character, each state can transform into every other with only a single step. Hence, a data set that consists only of unordered multi-state characters may only result in partial resolution of relationships among the study taxa. If each taxon has a unique code, there will be no resolution. If there are some shared codes, it will result in a bush, where each of the terminals are defined by a single state because such characters permit all possible character state transformations and no state will unequivocally diagnose a monophyletic group. Consequently, in a mixed data set of binary and unordered multistate characters, it is the binary characters that will be responsible for most or all of the resolution on the MPC. If the binary characters support a fully bifurcating MPC, then they will dictate all of the resolution. Any unordered multistate characters in the data set will simply be mapped parsimoniously onto that topology. Under these circumstances, the multistate characters add nothing to the analysis and may as well be excluded from the data set. However, if the binary characters permit only

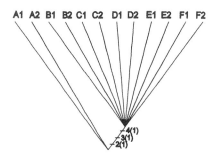

Figure 3.7 The strict consensus tree derived from the 45 equally most-parsimonious trees found from analysis of the three, unordered, multistate characters 2–4 (method B) using NONA. Only one clade is recovered (taxa B1–F2), supported by the presence of an FTC. However, this support is replicated three times and thus coding method B introduces redundancy into an analysis.

partial resolution of the MPC, then it is possible for an unordered multistate character to provide additional resolution. Such is the situation in the present example. The TCCs only partially resolve the relationships of the taxa A1–F2 (Fig. 3.4), and thus certain states of character 1 can lend support to additional groupings and improve resolution. Character states 1(3) and 1(4) provide ambiguous evidence for the monophyly of the groups D1+D2 and CEF respectively, while character state 1(2) provides unambiguous support for the group C1+C2.

Method B: three separate multistate characters (characters 2–4)

As for character 1, there is no justification for regarding the states of characters 2–4 as forming ordered sequences and so these characters were also treated as unordered. When characters 2–4 were analysed alone, Hennig86 found 58 MPCs, NONA found 45 and PAUP found 710. Each of these MPCs is seven steps long (CI = 0.85; RI = 0.90). The strict consensus tree of the NONA cladograms (Fig. 3.7) contains only a single informative component. This clade is based upon the presence of an FTC. However, there is a problem in that each of the changes from state 0 to state 1 in characters 2–4 represents the gain of an FTC. In other words, by replicating this item of information three times, redundant data have been introduced into the analysis and spurious support afforded to the group BCDEF. This redundancy is a major flaw in coding method B and shows that the three characters, 2–4, are not completely independent.

When the TCCs were introduced, Hennig86 found only two MPCs, NONA found one and PAUP found 31. The cladogram found by NONA (Fig. 3.8) is the strictly-supported cladogram and has a length of 12 steps (CI = 0.91, RI = 0.95). The second cladogram found by Hennig86 differs only in spuriously resolving B1 and B2 as a group. Again, the group BCDEF is supported by replicated redundant information, while a reversal from state 2 to state 1 in character 3 supports the monophyly of the clade C1+C2, resolution that is additional to that provided by the TCCs alone.

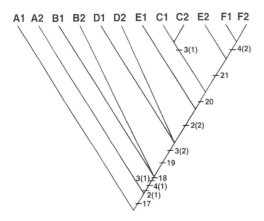

Figure 3.8 The single most-parsimonious and strictly-supported tree found from the analysis of the three, unordered, multistate characters 2–4 and the TCCs.

Thus coding the FTC data as three independent multistate characters, rather than as a single multistate, has improved resolution by one clade but that clade is over-supported by redundant replicated information.

Method C: contingent coding (characters 5–8)

Contingent coding differs from the previous two methods in treating the presence/absence of an FTC (character 5) as a separate character from those that code the variation observed within it (characters 6–8). Consequently, however, the latter three characters must be coded as non-applicable (i.e. as '?') in those taxa that are coded 0 for character 5 (i.e. FTC absent).

When characters 5–8 were analysed alone, Hennig86 found 58 MPCs, NONA 45 and PAUP 710. Each of these MPCs is five steps long (CI = 0.80; RI = 0.87). The strict consensus tree of the NONA cladograms (Fig. 3.9) again contains only a single informative component, based upon the presence of an FTC (character 5). However, in contrast to the previous analysis using coding method B, there is no replication of this information and no redundancy. This is reflected in a decrease in the length of the MPC from seven to five steps. In other words, by introducing question marks to accommodate non-applicable conditions, contingent coding has rendered the presence of the FTC independent of the three characters that code for variation within it.

As with coding method B, when the TCCs were included, Hennig86 found two MPCs, NONA a single MPC and PAUP 31 MPCs. The strictly-supported MPC found by NONA (Fig. 3.10) has the same topology and distribution of character information as the equivalent cladogram found using coding method B (Fig. 3.8), but without the redundancy in character information.

Thus contingent coding represents an improvement over treating the FTC data as three separate multistate characters in that no redundant character distributions are included on the MPC. However, the introduction of question marks brings its own problems (see above), although these are not manifest in this particular example.

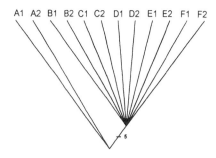

A1 A2 B1 B2 C1 C2 D1 D2 E1 E2 F1 F2

Figure 3.9 The strict consensus tree derived from the 45 equally most-parsimonious trees found from analysis of the four, contingent coded characters 5–8 (method C) using NONA. Only one informative component is recovered (taxa B1–F2), supported by the presence of an FTC. However, unlike the tree in Fig. 3.7, there is no replication of support for this clade and no redundancy is introduced into the analysis.

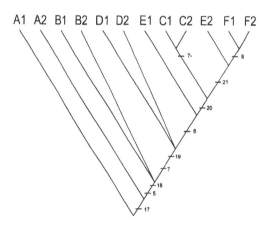

A1 A2 B1 B2 D1 D2 E1 C1 C2 E2 F1 F2

Figure 3.10 The single most-parsimonious and strictly-supported tree found from analysis of the four, contingent-coded characters 5–8 and the TCCs.

Method D: absence/presence (A/P) coding (characters 9–15)

As with contingent coding, A/P coding treats the presence/absence of an FTC (character 9) as a character separate from those that code the variation observed within it (characters 10–15). However, in contrast to the three methods considered so far, A/P coding treats the presence or absence of each observed condition of a structure as a separate character.

When characters 9–15 were analysed alone, Hennig86 and NONA both found two MPCs, while PAUP found 28. Each of the two MPCs found by Hennig86 and NONA is ten steps long (CI = 0.70; RI = 0.85). Both cladograms (Fig. 3.11A and B) are strictly-supported and almost fully resolved. The first MPC (Fig. 3.11A) divides the group BCDEF into two major clades, BC and DEF, supported by characters 12 and 13 respectively. Character 10 independently supports the groups

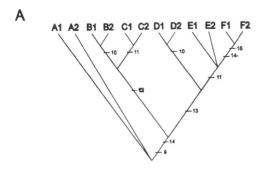

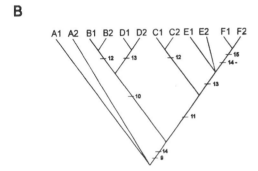

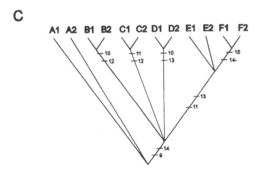

Figure 3.11   A,B. The two equally most-parsimonious and strictly-supported trees found by analysis
of the seven presence/absence coded characters 9–15 (method D) using NONA. C. The
strict consensus tree of the two trees in A and B.

B1+B2 and D1+D2, while character 11 provides similar support for the groups
C1+C2 and EF. On the second cladogram (Fig. 3.11B), the reverse pertains. This
time, group BCDEF is divided into clades, BD and CEF, which are supported by
characters 10 and 11 respectively. Character 12 now occurs independently in the
groups B1+B2 and C1+C2, and character 13 in groups D1+D2 and CEF. This
conflict is apparent in the strict consensus tree (Fig. 3.11C), in which the relationships of the groups B1+B2, C1+C2, D1+D2 and EF are unresolved.

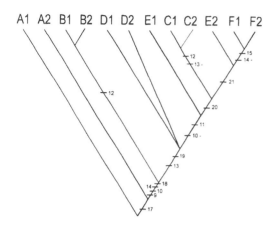

Figure 3.12 The single most-parsimonious and strictly-supported tree found from analysis of the seven presence/absence coded characters 9–15 and the TCCs.

When the TCCs were included, Hennig86 found two MPCs, NONA found a single MPC and PAUP found 33. The NONA cladogram (Fig. 3.12) was the strictly-supported topology (the second cladogram found by Hennig86 differed in grouping D1 and D2 together). The topology in Fig. 3.12, which has a length of 16 steps (CI = 0.75; RI = 0.87), is essentially that supported by the TCCs but with additional resolution of the groups B1+B2 and C1+C2, both on the basis of the presence in these taxa of an adpressed FTC (character 12).

It would thus seem that presence/absence coding offers improved resolution compared to that obtained using methods A–C. However, presence/absence coding suffers from major theoretical defects.

First, as we noted above, with the exception of the inclusive character, A/P characters form a set of logically dependent pairs. Allocating codes 0 and 1 to one character of such a pair logically implies the inverse coding of 1 and 0 for the other character. Thus, for example, those characters coded 1 in character 10 (short FTC) are automatically coded 0 for character 11 (long FTC) and vice versa. Thus the requirement that cladistic characters be logically independent is violated.

The FTC data illustrate another flaw in A/P coding, which we referred to above as the 'off-on' effect. In both MPCs (Figs 3.11A and B), the monophyly of clade BCDEF is supported by character 14, presence of a non-recurved FTC. However, this character is then subsequently reversed on the branch leading to F1+F2, at the same time as character 15, the presence of a non-recurved FTC, appears in support of this group. Thus a change in a single attribute of the FTC, namely the curvature, results in two steps on this branch, not one. Consequently, it would take three other characters to break up this group, rather than the expected two. A similar pattern can be observed in Fig. 3.12.

A third defect of A/P coding is exemplified by the difficulties faced in interpreting the optimizations of the A/P characters in terms of transformations in the observed features of the FTC as a whole. In Fig. 3.11A, when the FTC is developed initially (character 9), it also appears in a non-recurved form (character 14). However, we can say nothing at this point regarding either its shape (adpressed or looped), which only

Table 3.3 Sankoff matrix applied to character 16

	0	1	2	3	4	5
0	–	3	3	3	3	3
1	1	–	1	1	2	3
2	1	1	–	2	1	2
3	1	1	2	–	1	2
4	1	2	1	1	–	1
5	1	3	1	2	1	–

becomes evident on the branches leading to BC and DEF, or its length (short or long), which is only specified on the branches leading to B, C, D and EF. (The same problem occurs in the second MPC, Fig. 3.11B, where the order of development of shape and length is reversed, and a similar pattern is evident in Fig. 3.12.) Yet, when the FTC first appears, it must have a particular length, shape and curvature. Any other transformational interpretation is nonsense. It is possible that A/P coding could be justified under the taxic approach to cladistics (Eldredge, 1979), in which only the distributions of characters among taxa are used to hypothesize group membership. Be that as it may, the defect of logical dependency of presence/absence characters still totally undermines this approach to character coding.

Lastly, we tried Sankoff coding of a single multistate character (character 16). In the Sankoff matrix (Table 3.3), values in the upper triangle indicate the forward transformation costs between states (e.g. $0 \rightarrow 1$ or $3 \rightarrow 5$), while values in the lower triangle indicate the transformation costs in the reverse direction (e.g. $3 \rightarrow 1$ or $5 \rightarrow 4$). The matrix is asymmetrical in that the costs in the first line (gain of the FTC from the 'absent' condition) are not the same as those in the first column (loss of the FTC). In the present example, when the FTC is gained (as noted in the previous section), it must appear with some form of length, shape and curvature, i.e. three features. Hence each entry in the first row is a 3. However, we consider it appropriate to treat the loss of the FTC as a single step (i.e. when the FTC is lost, all features relating to it are necessarily lost simultaneously) and hence all the entries in the first column are 1. The remaining entries in the Sankoff matrix indicate how many of the three independent features of the FTC (length, shape, curvature) change in order to convert one observed form of the FTC to another. For example, to transform a short, adpressed, non-recurved FTC (state 1) into one that is long, looped and non-recurved (state 4), two features (length and shape) must change and thus the entry in the Sankoff matrix is 2. Currently, analysis of Sankoff coded characters can only be undertaken using PAUP.

Analysis of character 16 alone produced 622 MPCs, each with a length of 5 steps (CI = 1.00; RI = 1.00). The strict consensus tree (Fig. 3.13) contained only a single informative component, supported by the presence of an FTC (character 16, state 1). When the TCCs were added, then 27 MPCs were found, with a length of 10 steps (CI = 1.00; RI = 1.00). The strict consensus tree of these 27 cladograms (Fig. 3.14) also had a length of 11 and thus represents the strictly-supported cladogram for these data.

Overall, it is clear that there is actually very little grouping information contained within the FTC data when they are considered by themselves. Nevertheless, a

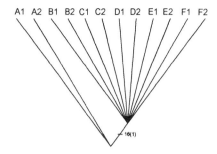

Figure 3.13 The strict consensus tree and strictly-supported tree derived from the 622 equally most-parsimonious trees found from analysis of the Sankoff-coded character 16 using PAUP. As for coding method D, only one informative component is recovered (taxa B1–F2), supported by the unreplicated presence of an FTC.

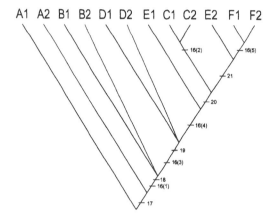

Figure 3.14 The strictly supported tree found from analysis of the Sankoff coded character 16 and the TCCs.

comparison of the results of coding methods A–D with those obtained from Sankoff coding demonstrates the superiority of the latter method in representing the variation observed in the FTC (Fig. 3.3). Coding the data as a single multistate character (method A) produces no resolution, whereas Sankoff coding recovers a single clade, BCDEF, supported by the presence of an FTC. Analysis of the multistate character after breaking it down into three separate characters does allow clade BCDEF to be recovered, but at the expense of including redundant, replicated information and the production of multiple non-combinable MPCs. Analysis of the Sankoff coded data yields only a single strictly-supported MPC, which is identical in topology and character data distribution to one of those derived from the three separate multistate characters (Fig. 3.6D). Contingent coding (method C) removes the redundancy from the data but introduces problems of its own due to the necessary inclusion of question marks to accommodate non-applicable character states. These are of no consequence in the FTC data set but their detrimental effects in other data sets have been well documented. The defects of A/P coding are such that this method must be rejected as a general coding method. It would therefore appear that the only justifiable coding method for characters that are absent in some taxa and display variation in form among others is Sankoff coding.

Conclusions

The multistate character is used in cladistic analysis where there is reason to believe that there is biological and/or logical dependency between individual observations that are coded as states within a character. The most frequent usage incorporates an assumption of ancestor–descendant relationships between states and as such agrees with a transformational rather than a taxic view of a character. There have been previous attempts to determine such transformational relationships through the analysis of character-state trees (Lipscomb, 1992), but these attempts have been concerned with state-adjacency and order (see discussion in Scotland and Williams, 1993).

Several methods have been used to code multistate characters. All have advantages and disadvantages but the important points are that, when used in conjunction with binary characters, they can lead to different systematic conclusions and different theories of transformational evolution. We review by example such effects and conclude that Sankoff coding, used as an attempt to quantify change between morphological variation, offers the most satisfactory method of coding where theories of character transformation are the desired outcome of cladistic analysis.

Acknowledgements

We thank Robert Scotland for inviting us to participate in the meeting which stimulated this volume. We both separately prepared talks which advocated presence/absence coding as the desideratum but both realized the shortcomings of that method while participating – the true value of a scientific meeting. We are also especially grateful to Ward Wheeler who reminded us of the little-used Sankoff coding.

Appendix 3.1: Caecilian morphological characters based on data given in Wake (1993)

```
                          IIIIIIIIII222222222233333
                          1234567890123456789012345678901234
A Epicrionops petersi     000000100110000000000000101121002 00
B Dermophis mexicanus     000000023100000111301002100020 02 10
C Gymnopis multiplicata   511100102213100111210002101020 02 10
D Caecilia occidentalis   401100223213100111210003011100 00 10
E Oscaecilia ochrocephala 411011034223201111310100001000 00 10
F Geotryptes seraphini    201100010102100111210002111020 00 10
G Hypogeophis rostratus   001100003101100111310000010100 00 10
H Idiocranium russelli    301101021101100111110003000100 00 10
I Scolepomorphus ulugurensis 5020110342232101111200221112200110
J Typhlonectes natans     101000103010101111010012110000 11 11
                          *  *     ******     **  **    **  *
```

Appendix 3.2: Data in Appendix 3.1 recoded as presence/absence

```
          1111111111222222222233333333334444444444555555555566666666667777777
 1234567890123456789012345678901234567890123456789012345678901234567890123 4567
A 111110010000000101000100000100101000000001001100001001110011010100 0000100
B 111110010000001000100000100101001001001000011100010011010100100010000 0110
C 111111010100010101000010000010100011001001000100010101100001000001 0010
D 001100000101000010010000010000010000100010100100100101101000010001000 0110
E 001100010011100000100000100001000001001000101011001011010100000100010010
F 110101000101001000100010000010000010101111000010010111100001001 0010
G 111111000101001000100100000100000010001100010001010111100001000100 10010
H 011100000101011000010010000010000010001100010100010011100001010000 10010
I 000000000001011000000100000010100010010000100100101100100101011100100 01010
J 111111100100000010100000010100010000100000010011110001000000011101011001011
```

Data recoded as presence/absence using a combination of the methods D1 and D2 from Table 3.1. Characters 1–39, eye characters; 40–56, ear characters; 56–67, hypoglossal nerve characters; 68–77, olfactory/vomero-nasal characters.

Wake's original coding incorporating many multistate characters (asterisks). Characters 1–13, eye characters; 14–23, ear characters; 24–27 hypoglossal nerve characters; 28–34, olfactory/vomero-nasal characters.

References

Archie, J.W. (1985) Methods for coding variable morphological features for numerical taxonomic analysis, *Systematic Zoology*, **34**, 326–345.

Bryant, H.N. (1989) An evaluation of cladistic and character analyses as hypothetico-deductive procedures, and the consequences for character weighting, *Systematic Zoology*, **38**, 214–227.

De Pinna, M.C.C. (1991) Concepts and tests of homology in the cladistic paradigm, *Cladistics*, **7**, 367–394.

Eldredge, N. (1979) Alternative approaches to evolutionary theory, *Bulletin of the Carnegie Museum of Natural History*, **13**, 7–19.

Farris, J.S. (1988) *Hennig86 Version 1.5. MS-DOS Program*, Port Jefferson Station, New York: published by the author.

Gardiner, B.G. (1982) Tetrapod classification, *Zoological Journal of the Linnean Society of London*, **74**, 207–232.

Gardiner, B.G. (1993) Haematothermia: warm-blooded vertebrates, *Cladistics*, **9**, 369–395.

Gauthier, J., Kluge, A.G. and Rowe, T. (1988) Amniote phylogeny and the importance of fossils, *Cladistics*, **4**, 105–209.

Goloboff, P. (1996) *NONA Version 1.50. MS-DOS Program*, San Miguel de Tucumán, Argentina: published by the author.

Hawkins, J.A., Hughes, C.E. and Scotland, R.W. (1997) Primary homology assessment, characters and character states, *Cladistics*, **13**, 275–283.

Lipscomb, D.L. (1992) Parsimony, homology and the analysis of multistate characters, *Cladistics*, **8**, 45–65.

Maddison, W.P. (1994) Missing data versus missing characters in phylogenetic analysis, *Systematic Biology*, **42**, 576–581.

Mayr, E., Linsley, E.G. and Usinger, R.L. (1953) *Methods and Principles of Systematic Zoology*, New York: McGraw-Hill.

Nelson, G.J. and Patterson, C. (1993) Cladistics, sociology and success: a comment on Donoghue's critique of David Hull, *Biology and Philosophy*, **8**, 441–443.

Nixon, K.C. and Carpenter, J.M. (1996) On consensus, collapsibility, and clade concordance, *Cladistics*, **12**, 305–321.

Patterson, C. (1982) Morphological characters and homology, in Joysey, K.A. and Friday, A.E. (eds) *Problems in Phylogenetic Reconstruction*, London: Academic Press, pp. 21–74.

Pimentel, R.A. and Riggins, R. (1987) The nature of cladistic data, *Cladistics*, **3**, 275–289.

Pleijel, F. (1995) On character coding for phylogeny reconstruction, *Cladistics*, **11**, 309–315.

Pogue, M.G. and Mickevich, M.F. (1990) Character definitions and character state delineation: the *bête noire* of phylogenetic inference, *Cladistics*, **6**, 319–361.

Scotland, R.W. and Williams, D.M. (1993) Multistate characters and cladograms: when are two stamens more similar to three than to four? A reply to Lipscomb, *Cladistics*, **9**, 343–350.

Smith, A.B. (1994) *Systematics and the Fossil Record: Documenting Evolutionary Pathways*, Oxford: Blackwell Scientific.

Swofford, D.L. (1993) *PAUP, Phylogenetic Analysis Using Parsimony, Version 3.1. Macintosh OS Program*, Washington, DC: Smithsonian Institution.

Thiele, K. (1993) The holy grail of the perfect character: the cladistic treatment of morphometric data, *Cladistics*, **9**, 275–304.

Wake, M.H. (1993) Non-traditional characters in the assessment of caecilian phylogenetic relationships, *Herpetological Monographs*, no. 7, 42–55.

Wake, M.H. (1994) The use of unconventional morphological characters in analysis of systematic patterns and evolutionary processes, in Grande, L. and Rieppel, O. (eds) *Interpreting the Hierarchy of Nature*, San Diego: Academic Press, pp. 173–200.

Wiley, E.O. (1981) *Phylogenetics: the Theory and Practice of Phylogenetic Systematics*, New York: Wiley Interscience.

Wilkinson, M. (1995) A comparison of two methods of character construction, *Cladistics*, **11**, 297–308.

Wilkinson, M. (1997) Characters, congruence and quality: a study of neuroanatomical and traditional data in caecilian phylogeny, *Biological Reviews*, **72**, 423–470.

Chapter 4

On characters and character states: do overlapping and non-overlapping variation, morphology and molecules all yield data of the same value?

Peter F. Stevens

Introduction

> Trichosporeae are still defined only by their appendaged seeds; yet the boundary between seeds with fine points and seeds with appendages will be seen differently by different workers'
>
> (Burtt, 1997: 87)

> Continuously varying characters are not the stuff of phylogeny.
>
> (Crowe, 1994: p. 78).

Patterson and Johnson (1997: 361) observed that 'the emphasis has shifted from observation, the source of the matrix, to whatever message can be extracted from the matrix. . . . This change of emphasis replaces our pernicious old black box, evolutionary systematics, with a new one, the matrix.' Although they suggested that this shift was due to the general acceptance of the cladistic approach, the same emphasis on manipulation of data pervades numerical phenetics (e.g. Sokal and Sneath, 1963; Sneath and Sokal, 1973: see Stevens, 1991). Indeed, it is perhaps almost inevitable, given the nature and ever-increasing power of computers. Yet Patterson and Johnson (1997) showed how an error rate of 11% in osmeroid fish data had substantial effects on the topology resulting from analysis of those data (see also Raikow *et al.*, 1990; Kesner, 1994), and pleaded for more care in the compilation of data matrixes. Pheneticists had similar concerns, evaluating how scrambling states of different numbers of characters affected the topologies of the cladograms produced after analysing those data (Fisher and Rohlf, 1969).

But how do we convert the variation that we so laboriously measure into the 0s and 1s of a data matrix (Pimentel and Riggins, 1987; Gift and Stevens, 1997; Swiderski *et al.*, 1998)? What are the implications for subsequent analyses of what we do? Many such morphological data – data-III in the sense of Stevens (1996), the 0s and 1s of a data matrix – are quantitative characters in which variation, whether continuous or meristic, is overlapping. Such overlapping variation does not yield character states that can be delimited unambiguously, and different observers delimit states in the same measurements differently (Gift and Stevens, 1997). Yet measurements of individuals (data-I), or even ranges or some summary of measurements of the taxa being studied (data-II), are rarely if ever presented in a systematic study (Stevens, 1996). In some morphological studies in particular the whole process is so problematical that subsequent cladistic analyses of the data are compromised

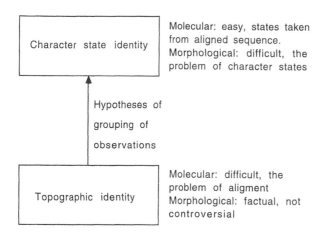

Figure 4.1 Steps leading to the production of character states as suggested by Brower and Schawaroch (1996).

(Gift and Stevens, 1997); another less serious consequence is persistent terminological confusion (e.g. Patterson, 1988).

This is the black box, and we need to find out what goes on inside it. First, I outline a framework to facilitate discussion about the various operations involved in delimiting characters and deciding what their states are. I build on recent clarification of de Pinna's (1991) concept of primary homology by Brower and Schawaroch (1996: see Fig. 4.1) and a comparison of data used in morphological and molecular phylogenetic studies (Stevens, 1996) to provide a schematic of the main steps involved in morphological and molecular studies (Fig. 4.2). Then I report on phylogenetic analyses of different kinds of data-III obtained from the same group of organisms. Some were obtained from characters in which variation is overlapping and the states delimited following visual inspection of the variation pattern. Other data-III are derived from variation in which the basic pattern is non-overlapping.

First, a few definitions. A character is any feature shared among organisms that we think will provide information to use in phylogenetic analysis (Colless, 1985; Fristrup, 1992, for reviews). Several authors consider characters and states to be indistinguishable, differing only in the hierarchical level at which they are applicable (e.g. Bock, 1974; Platnick, 1979; Patterson, 1988). Even if appropriate when discussing trees (but cf. Hawkins et al., 1997), when making observations on organisms, it is conceptually useful to distinguish between character, as defined above, and its states. In practice, a character is the sum of features showing particular similarities (e.g. Patterson, 1982; Stevens, 1984), topographical homologies (Jardine, 1969), topographical identities (e.g. Brower and Schawaroch, 1996), or relationships of primary homology (e.g. de Pinna, 1991, Fig. 1: the synonymy is lengthening) with each other that we observe in different organisms. We group the various manifestations of a character into states to make the character of use in phylogenetic studies.

Many issues concerning characters, states, and their use in phylogenetic studies are ignored here. These include whether a character should be single, with many states, or subdivided into independent binary characters, or whether characters are

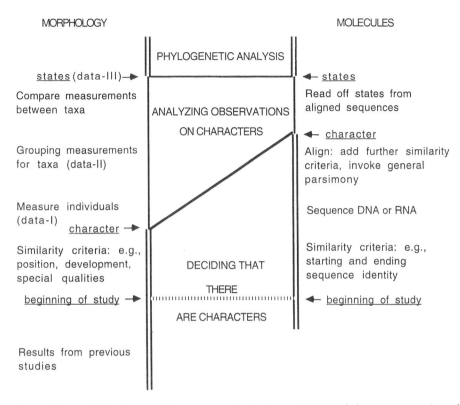

MORPHOLOGY   MOLECULES

PHYLOGENETIC ANALYSIS

states (data-III) →   ← states

Compare measurements   Read off states from
between taxa   ANALYZING OBSERVATIONS   aligned sequences
                                    ON CHARACTERS   ← character

Grouping measurements   Align: add further similarity
for taxa (data-II)   criteria, invoke general
                                                           parsimony

Measure individuals   Sequence DNA or RNA
(data-I)
        character →

Similarity criteria: e.g.,   Similarity criteria: e.g.,
position, development,   DECIDING THAT   starting and ending
special qualities   sequence identity
                                    THERE
beginning of study →   ← beginning of study

                                ARE CHARACTERS

Results from previous
studies

Figure 4.2 Amplified schema of the steps leading to the production of character states in molecular and morphological studies.

linked either logically or biologically, so that what appear to be several characters should all be treated as one (e.g. Wake, 1994; Fink and Zelditch, 1995; Pleijel, 1995; Wilkinson, 1995; Kellogg and Juliano, 1997; Hawkins *et al.*, 1997; Doyle, 1997). Having decided on characters and their states, issues of scaling and weighting arise (e.g. Farris, 1990; Wheeler, 1995), the latter in molecular studies when transitions, transversions, insertions and deletions are differentially weighted, and also in morphological studies when states in multistate characters can be ordered or unordered. What homology 'is' – synapomorphy (e.g. Patterson, 1982; Stevens, 1984), secondary homology (de Pinna, 1991 and references therein), or something else – remains a matter of debate (see Hall, 1994); I find the word so ambiguous that it is best replaced by the synonyms of its particular usages. Finally, how structures we call homologies or synapomorphies have arisen during evolution, and how characters are individuated (e.g. Wagner, 1989, 1994), are not discussed here.

Characters and character states in morphological and molecular studies

The sequence of operations involved in choosing characters and delimiting states is outlined in Fig. 4.2. Below I compare the main steps in molecular and morphological studies. Although the whole process is presented as if it were linear, this is for illustrative purposes only.

Specification of similarity or topographic identity

In all studies an initial decision is made that the same structure or sequence is being compared in different organisms, a decision based on the fundamental, but rather elusive, concept of similarity. Similarity can be thought of as a series of 1:1 correspondences between objects (e.g. Woodger, 1945; Patterson, 1988). Most criteria by which similarity is judged are similar to those suggested by Remane (1952) – his criteria of homology. These criteria are in turn based on the principles enunciated by early workers such as Etienne Geoffroy Saint-Hilaire (see Rieppel, 1988, esp. pp. 37–39) and Richard Owen (see Boyden, 1943; Panchen, 1994). Such criteria are used in molecular studies (e.g. Brower and Schawaroch, 1996), but the role they play there has occasioned considerable (e.g. Patterson, 1988; Mindell, 1991b; Donoghue and Sanderson, 1994), but, I argue below, sometimes rather misdirected, discussion.

Remane's (1952) three main criteria of homology (= similarity) are similarity in topographical relationships, similarity in special structure, and continuity of form established by intermediates, whether during development or in different adult forms (one of his secondary criteria suggests the later step of phylogenetic analysis – e.g. Donoghue and Sanderson (1994)). These criteria are in large part but not entirely topographical, and help to specify points of comparison or identity between members of the group being studied (Riedl, 1976; Rieppel, 1988). Brower and Schawaroch (1996) note that the alignment of sequences, the step of determining similarity (their topographic identity: see Fig. 4.1) in molecular studies, is controversial. They consider the comparable step in morphological studies to be largely non-controversial, referring here to de Pinna, who observed 'Similarity or topographical correspondence [Brower and Schawaroch's topographical identity] is factual' (de Pinna, 1991: 373); Patterson (1988) also thought that similarity could be factual.

Molecular studies

Molecular studies – and here I restrict myself to polymerase chain reaction (PCR) sequencing studies – initially use similarity criteria to specify what is to be compared in different organisms. Amplification primers are shorter or longer sequences which need to match completely with sequences in the target genomes if any data are to be obtained; they control what within the genome will be sampled by specifying the place where a PCR starts. This complete match is a 1:1 correspondence specified by particular bases (Remane's special properties) in a particular sequence (topographic criterion). Note, however, that PCR practically never results in the discovery of a single character to be used in subsequent phylogenetic analysis. An area within which characters will subsequently be looked for by the systematist has been pinpointed by the amplification primers.

If all members of the group have DNA sequences of the same length that are largely similar, then the next step is the determination of character states (see below). If, however, the sequences are of different lengths, or are very dissimilar, then alignment, a complex set of operations (Davison, 1985; Waterman *et al.*, 1991; Wheeler, 1996; Hillis *et al.*, 1996, for references; also see Chapter 5), clarifies the positional relationships between bases on different sequences, restoring their 'positional homology' (Moritz and Hillis, 1996). In the simplest form of this alignment process, gaps are added or removed in a second application of similarity criteria, one that finally sets the topological or similarity relationships between the bases – the individual characters – on each sequence (of course, the whole sequence can be treated as one character in combined analyses with other data, as by Doyle (1992, 1997)). The goal is to minimize differences (or maximize the identity) between sequences, whether prior to or during (Mindell, 1991a; Wheeler and Gladstein, 1992) phylogeny inference. Parsimony-based comparisons can decide which similarity scheme yields the greatest summed identities between the sequences. This is indeed a controversial step, as Brower and Schawaroch (1996) note. In subsequent phylogenetic analyses the states obtained can be used with the same weightings to determine the phylogeny.

Molecular characters are very largely established by positional criteria. When a base in a particular position changes, there is no information in the replacing base that marks it as being the same character (conversely, the guanine molecules in, say, positions 123, 346, and 348 are indistinguishable). There is no site-specific identity between the different bases at the one site, and the observer infers that they belong to the same character only because the majority of bases elsewhere in the sequence show substantial identity across the taxa being examined, an issue to which I will return.

The role of intermediates (Remane's third criterion) is interesting. If molecular sequences are difficult to align, sequencing taxa with intermediate sequences may be important. To help in the alignment process, taxa can be inputted in order of decreasing phylogenetic relationship (e.g. Mindell, 1991a, 1991b), or most similar sequences successively compared, as in the CLUSTAL algorithm. The addition of intermediate sequences can affect very large numbers of characters.

Morphological studies

In contrast to molecular studies, the complex iterative process by which the identity of characters is determined during the alignment process seems not to take place in morphological studies. Yet although morphological similarity is often considered to be factual, it may be decidedly more problematical, as de Pinna (1991: 377) himself suggested when he noted that it was a vague notion that was a primitive concept in systematics (see also Rieppel, 1988; Smith, 1994). Even our simple belief that two organisms are related may affect what we see – we may expect that both will show the same or similar features (Norman *et al.*, 1992). We may not fall into this trap frequently, but if our character state definitions are imprecise, then it is easy to think of bark being fibrous, or a style being long, if we think that a specimen we are looking at belongs to a taxon with 'fibrous' bark or a 'long' style. It is certainly a mistake I make when making preliminary identifications of recently-collected material – especially when I am tired!

Similarity in general is a very elusive concept, perhaps rather more than we would like to admit (e.g. Goodman, 1972; de Pinna, 1991; Medin *et al.*, 1993). The length

Table 4.1 Interpretation of the androecial structure in *Decaphalangium*

	As fascicles of stamens	As stamens
Stamen number	Hundreds	10
Stamens fasciculate?	Yes	No
Anther dehiscence	Apical	Introrse
Anthers locellate?	No	Yes
Anther size	<0.4 mm	2.5–3 mm

of a leaf is readily enough specifiable, the apex and the base being recognizable reference points. But other apparently equally simple characters run into problems. What about the length of the lamina, and especially its width? The actual base of the lamina may be difficult to ascertain if it is acute or decurrent on the petiole, while lamina width may well refer to morphologically non-equivalent points in different members of the study group. Unless otherwise qualified, it is simply the width of the lamina at its widest point, wherever that happens to be – this is hardly satisfactory (McLellan and Endler, 1998, for references).

Furthermore, two observers may agree as to what they are seeing, but may disagree how what they see is to be interpreted. Similarity assessments may be dependent on theory of some sort (Jardine, 1969; Medin *et al.*, 1993), and a process analogous to the alignment of sequences. How are digits in a limb to be compared (Rieppel, 1994)? Are particular fossils to be compared with echinoderms or chordates (Smith, 1994: 34; see also Rieppel 1988: 44–49)? Similarly, the gonophyll theory (e.g. Melville, 1983), the telome theory (Zimmermann, 1965), and the anthocorm theory (Meeuse, 1975, 1984), to name just three, radically affect what characters we see in the same plants. At a lower taxonomic level, *Decaphalangium peruvianum* was described by Melchior (1930) as having an androecium consisting of ten fascicles, or aggregations of stamens – unique in Clusiaceae, to which it belongs. But are those ten fascicles better interpreted as being ten rather complex stamens? The scoring of five or more characters is affected by differing interpretations of what two observers would agree was at one level the same part of the organism (Table 4.1). Thus different and even conflicting sets of characters can be used to describe the same organism. But the cladistic consequences of this can be explored, much as Sattler and Rutishauser (1990) suggested for their complementary descriptions of plant form in which the conventional atomisation of the plant body is replaced by alternatives.

These examples suggest that although the decision that a morphological character may contain cladistic information can be made early in a study and seems unproblematical, this misrepresents the nature of the decisions made. Solutions to many morphological questions have been hammered out by discussions between biologists over the past two centuries (see also Walters, 1964; Inglis, 1966), and what has been a largely topographic alignment process (Jardine, 1967) now allows us to make almost unconscious decisions about the circumscriptions of characters. Nevertheless, the delimitation of characters is continually being questioned. What was initially considered to be a separate character may turn out to be the same as another character, or better subdivided into two or more characters. Similarly, in morphometric work in which complex shape transformations are analysed, characters become apparent only towards the end of the study (Zelditch *et al.*, 1995).

Intermediates, whether in different organisms, or uncovered during developmental work, play a similar role to those in a molecular study. They may help confirm that a structure in one organism is similar to an apparently rather different structure in another organism. In the case of *Decaphalangium*, the discovery of taxa with intermediate morphologies between the androecium of *Decaphalangium* and the more conventional stamens of *Clusia*, to which *Decaphalangium* probably belongs (Stevens, 1999), may clarify the similarity relationships of the former.

The delimitation of character states

Molecular studies

The delimitation of character states is a simple task. Nucleotide variation is discrete, so variation at a particular site can readily be recorded as states, and a distinction is often made between transitions, tranversions and indels, so the states can be weighted. There is no problem with intermediates, since molecular characters are discrete variables; at most, there may be polymorphism at the same site. Indeed, if at the very beginning of a study two sequences have a particular number of variable characters, there will be no fewer variable characters at the end, no matter how many taxa are added, providing the alignment is unchanged. Character states will not need to be re-evaluated unless the expanded study suggests a different alignment. Both characters and states may then change, since a G in one sequence may now be aligned with an A in position 50, rather than with a G in position 65 in the initial comparison.

Morphological studies

In morphological characters individual measurements or observations rarely convert to character states and the characters themselves are often quantitative (Stevens, 1991). However, there are no accepted procedures as to how individual measurements should be summarized, whether into bar graphs, box plots, etc., nor how data should be displayed or compared or states finally abstracted from data (Stevens, 1991, 1996; Thiele, 1994; Strait *et al.*, 1996; Gift and Stevens, 1997; Swiderski *et al.*, 1998). Gaps between states are often considered to be necessary, but the decision as to what might be a gap is no easy one to make – indeed, such decisions are often best described as cognitive–psychological, involving as they do poorly understood and very individualistic decisions that particular groups of observations represent characters (Gift and Stevens, 1997). Other ways of grouping observations into states are statistical in nature (Stevens, 1991; Thiele, 1994; Strait *et al.*, 1996; Swiderski *et al.*, 1998, for literature), although the procedures involved are often arbitrary and many need more observations than can be made on taxa represented by only a few specimens.

Even the use of strict absence/presence coding for characters (Pleijel, 1995) faces similar problems of delimitation – although of characters, not of states. Multistate characters can be reduced to a series of nominal variables, but Pleijel's (1995) hope that he had thereby avoided the problem of character state delimitation is unfortunately unfounded.

Intermediates – whether measurements or complex morphologies – play an ambiguous role in this stage of morphological studies. As we have seen, they may help confirm that a structure in one organism is similar to that in another. Indeed, an assumption is often made that there is a kind of continuity-in-principle between the different morphologies included in a character (see below). But the more that observations form a continuum of variation within a character, the more difficult the delimitation of states becomes, at least with many methods used (e.g. Gift and Stevens, 1997). What may seem to be a useful character at the beginning of the study may turn out to be worthless at the end. As a study expands, character states may have to be re-evaluated (Stevens, 1996), since additional observations may affect the initial delimitation of states. In this context, the relevance to cladistics of alternative approaches to understanding form that emphasize continuity between structures that are usually considered to be distinct, e.g. stem and leaf in plants (e.g. Sattler, 1992; Hay and Mabberley, 1994, and references therein), needs to be explored (e.g. see Chapter 7).

Discussion

Similarity criteria may be invoked twice in molecular studies as character states are abstracted from observation of variation (Fig. 4.2). Decisions about characters may not occur until the second application of similarity criteria, since individual characters *per se* are not initially of interest, rather, it is the sequence in its entirety. Characters are delimited (= aligned) with respect to the effect that alignment has on all other characters; the main problem in molecular studies is the recognition of characters, but not that of their states. In many morphological studies it is variation within the character, the delimitation of states, that is the main problem (see also Brower and Schawaroch, 1996), and individual characters are frequently recognised towards the beginning of a study.

However, some of the differences between molecular and morphological studies disappear on taking a broader perspective. Understanding the structure of organisms is in considerable part dependent upon understanding the topological relationships between the parts of organisms (Rieppel (1988, Fig. 4) implies, perhaps inadvertently, that topographical correspondence is invoked only after characters have passed the test of similarity, that is, after it has been decided that features in different organisms are similar to one another). This topographic alignment process has taken 200 or more years; furthermore, as discussed above, there is a morphological equivalent of changing sequence alignment and seeing what effect this has on characters. At the same time, prior knowledge of coding of amino acids and the structure and functioning of enzymes and messenger RNA may all affect how sequences are aligned (e.g. Kellogg and Juliano, 1997), hence there is also a 'historical' component in molecular studies. By 'aligning' the morphology of the organism or the base pairs of the sequence, the systematist ensures that each character is the same in all the organisms being studied.

Even if many morphological studies focus on character states, while in molecular studies the detection of characters is the immediate problem, the topographic criterion of similarity is paramount when delimiting both morphological and espe-cially molecular characters.

This clarification of how systematists decide upon characters and states bears on several areas of discussion.

1 The problem suggested by Donoghue (1992, p. 172), as he wondered 'when homology is defined in terms of similarity . . . how one set of similarity criteria can be justified as better than another', has a fairly direct answer in a molecular context. Alignment algorithms evaluate the consequences of changing structures in terms of the impact each change has on the interpretation of the whole structure, appeal being made to parsimony considerations before phylogenetic analysis of the data. The algorithms can take account of a gap of a single base pair in length, hence their power. This does not make this step of the process less problematic, but at least particular alignments are justified.

This option is not so readily available to the morphological systematist. In the case of *Decaphalangium*, mentioned above, separate analyses can be run using the differing interpretations of the staminal structures. Justifying the use of a partic-ular set of similarity criteria can then happen when the alternative cladograms are compared, both with each other and with cladograms produced using other evidence. However, morphological similarity relationships within an entire organism are rarely questioned in the same way as those within a particular sequence routinely are; reassessments of the morphologies of the whole organism are neither so easy to carry out, nor can their consequences be evaluated so readily. Morphologists rarely experiment explicitly with character delimitation.

2 A distinction has been drawn between taxic and transformational approaches in systematics. The transformational approach, with its emphasis on mechanism and process of change, not on the entities that are changing (Eldredge, 1979), has been considered vacuous; transformational similarities do not predict or specify groups, and are immune to testing by conjunction or congruence (Patterson, 1982). There is, however, another level at which transformation is at least implicitly invoked. One commonly 'sees' – or at least can imagine – transformation between different manifestations of the one character in different organisms (Brady, 1994; see also Hawkins *et al.*, 1997, and Chapter 2). Pimentel and Riggins (1987) imply as much when they suggest that the 'multistate variable representation [of a char-acter] corresponds to a transformation series'. Indeed, if we did not allow the possibility of such transformations, there would be no reason to think that we were looking at the same character. These transformations need have no particular direc-tion, but they serve conceptually to bind the different manifestations of the one character together; indeed, the more the character is a continuum, the easier such transformations appear (cf. Thom, 1992 and Chapter 7). The small gradations between members of the series allow topological transformations between them (Brady, 1994), so suggesting a fundamental commonality. In molecular studies, an inability to align sequences is equivalent to a failure to see how one sequence can transform into another.

3 The relative complexity of characters is often discussed (e.g. Patterson, 1988; Donoghue and Sanderson, 1994; Doyle, 1997). Here, as elsewhere in biology, complexity is not easy to evaluate (McShea, 1996), so when Patterson (1988) and Donoghue and Sanderson (1994) disagree over the relative complexity of morpho-

logical and molecular characters, the nature of their disagreement is not immediately obvious. Patterson (1988) suggested that in morphological studies, congruence was the final arbiter of homology (= synapomorphy). In molecular studies, on the other hand, homology rested in large part on the decision that two sequences were similar – to Patterson, a problem more of statistics than of common ancestry. Since statistical tests could not be applied to morphological characters, such characters seemed to him to be simpler: 'the similarity test in morphology is a weak one' (Patterson, 1988: 605; see also Bock, 1977, but cf. Doyle, 1997). Donoghue and Sanderson (1994), on the other hand, suggested that morphological characters might indeed be as complex as molecular characters.

However, complexity is often not a feature of individual characters as they are scored, but of the mostly topological relationships between such characters and others that are invoked when deciding that there are characters worthy of study. It is these relationships that impart complexity, and if there are differences in complexity between the two classes of characters, it may come from differences in these relationships. These differences may in turn relate to the developmental relationships of the characters within the organism (McShea, 1996). There are three main points bearing on this issue.

Firstly, the complexity of an individual character is not easy to evaluate. The variation encompassed by a molecular character is four bases, that of many morphometric characters, a series of measurements. A particular base appears in many different places in a sequence; many features of an organism may be 1.75 mm long (one could perhaps argue that although two thymine molecules are made up of the same numbers of the same kinds of atoms, two measurements of 1.75 mm long are potentially different, both being approximations). In both cases, the characters are very simple, and nothing in an individual measurement or base allows us to specify the character we have been measuring or analysing. Nevertheless, the changes in a bone or leaf blade that lead to these differences in length may be more complex than those leading to base variation at a particular position in a sequence (although the initial cause of those changes may be relatively simple).

A distinction is often made between qualitative and quantitative morphological characters (but see below). In terms of the variation recorded by the systematist, it can be argued that the former, features such as shape and texture, are likely to be more complex, even if they can be decomposed into a number of quantitative variables (Stevens, 1991; note that Fitter (1995) found that qualitative characters were more common at higher hierarchical levels, quantitative characters less common; characters of different complexity may vary in a similar fashion).

Secondly, discussion of similarity and complexity as attributes of characters may confuse levels. Molecular characters, nucleotide positions, are very simple compared to the complexity of the tissues that make up morphological characters. Furthermore, in direct sequence comparison, these positions show only very simple, topographic relationships, those between two adjacent bases. Yet if individual bases have only two bases adjacent to them, so at that level are far less complex than a skull bone with its often multiple relationships to other bones, an individual base is assigned a particular position because all the other bases of the sequence are simultaneously aligned. Hence the power of molecular similarity assessments, and why a molecular 'character' – when treated as an entire DNA sequence – seems complex.

Of course, whole sequences are different both from individual base positions and from morphological features. Each sequence may contain tens, hundreds, or even thousands of units, each one of which can be compared with units in other sequences. The 'statistical versions of similarity testing employed with molecular sequence data' (Patterson, 1988: 610) used to clarify the 'homology' of such sequences refers to the sequence as a whole – and not the units, the individual positions that provide the characters. Nevertheless, a statement that sequences are or are not homologous (= similar) could be based on a test for significance of the observed similarities between them.

But topographic relationships of an individual morphological character may be complex. A character is identified within an animal by relating it to various other features – for example, identifying a particular bone in terms of its positional relationship to other bones, the hole that signifies the passage of a particular nerve, and places where it is evident that muscles are attached. These muscles, nerves, and adjacent bones themselves show similar sets of relationships to yet other features, and so on. Thus the complexity of the manipulations allowed by the simplicity of a nucleotide sequence and that lead to a particular delimitation of a molecular character, still may not allow a claim that molecular characters are more complex than morphological ones. But, as mentioned above, morphological equivalents of different sequence alignments are at best very hard to do, they are rarely carried out, and they are not evaluated statistically in the way that alignments are.

There is a final wrinkle on this issue of complexity. If a gene, or region of DNA that assorts independently, is treated as a single character (Doyle, 1992, 1997), the characters in the discussion above become character states. If the complexity of a character is measured by the number of states into which it can be divided, such molecular characters show far greater complexity than that of any morphological character of which I am aware.

Thirdly, congruence is the ultimate arbiter in both molecular and morphological studies. Mindell (1991a) pointed out that alignment programmes resulting in 'homology assignments' (= similarity assessments) were based on quantitative, probabilistic procedures, but in this they were no different from many other aspects of systematic process. Homology (= synapomorphy) decisions are ultimately determined by the global manipulations of sequence similarity carried out during sequence alignment (Patterson, 1988) and the congruence tests of phylogenetic analysis (e.g. Miyamoto and Cracraft, 1991; Donoghue and Sanderson, 1994). This is true even of those procedures that have utilized phylogenetic relationships in seeking the optimal alignment; even there, there can be homoplasy.

4 Cladograms produced by different kinds of data are often compared (e.g. Sanderson and Donoghue, 1996, and references therein), or different kinds of data are analysed separately and then in combination in one simultaneous analysis. Distinctions are usually drawn between molecular and morphological data, and within the latter, developmental data are often singled out for attention. But another way of looking at data comes not from where in the organism, or on what aspect of it, the observations were made, but from the nature of the variation that was subdivided into states, e.g. whether variation is overlapping or non-overlapping. Do analyses of different kinds of morphological data produce cladograms with similar topologies and similar measures of support?

Table 4.2   Taxa included in the analysis (abbreviations following the names are those used in Table 4.4: *Rhododendron lapponicum* is not included in any group there)

Kalmia angustifolia (K. angus)
Kalmia ericoides (K. eric)
Kalmia cuneifolia (K. cune)
Kalmia hirsuta (K. hirs)
Kalmia latifolia (K. latif)
Kalmia microphylla (K. micro)
Kalmia polifolia (K. polif)
Loiseleuria procumbens (Lois)
Leiophyllum buxifolium (Leio)
Rhododendron lapponicum
Rhodothamnus chamaecistus (R'tham)
Phyllodoce coerulea (P. coer)
Phyllodoce empetriformis (P. empet)
Kalmiopsis leachiana (K'opsis)

Comparative analysis of data based on overlapping and non-overlapping morphological variation

Given the uncertainty surrounding the delimitation of character states when variation is overlapping, one obvious question to ask is, how do topologies resulting from the analysis of such states compare with those produced by the analysis of other kinds of variation? I address this question here in a comparative sudy of species of *Kalmia* (Ericaceae) and its relatives.

Methods

Taxa

The taxa are seven species of *Kalmia*, the monotypic genera *Loiseleuria* and *Leiophyllum*, both believed on various grounds to be derived from within *Kalmia* (Kron and King, 1996; Gift, Kron and Stevens, in preparation), and five outgroup taxa, *Rhododendron lapponicum*, *Rhodothamnus chamaecistus*, *Phyllodoce coerulea*, *P. empetriformis*, and *Kalmiopsis leachiana*. The taxa are listed in Table 4.2.

Characters

Quantitative characters in which there was overlapping variation (Table 4.3) yielded one group of matrices (O – overlapping), and characters in which there were gaps in the variation, or could be otherwise unambiguously divided into states (Table 4.4), yielded another group of matrices (N – non-overlapping).

For ten characters in which variation was overlapping, measurements were taken from between 10 and 50 individuals, depending largely on limitations of available material. Measurements were graphed showing 95% confidence intervals on the mean with either linear or $\log_{10}$ scaling of the ordinate, and also with bars showing 2 x SD of the mean. An example of the data collected for one character, petiole length, is shown in Fig. 4.3. Subjects in an experimental study were asked to delimit

Table 4.3 Characters showing overlapping variation used in the analyses (the cited figures appear in Gift and Stevens, 1997)

	O_a	O_b	O_c	O_d
Petiole length (Fig. 2a)	1	5	3	1
Lamina length (Fig. 2b)	1	6	2	1
Lamina width (Fig. 2c)	1	5	1	1
Internode length (Fig. 2d)	1	2	2	2
Pedicel length (Fig. 2e)	1	1	1	1
Sepal width (Fig. 2f)	4	7	7	7
Sepal length (Fig. 2g)	1	1	1	1
Corolla width (Fig. 2h)	4	9	3	9
Style length (Fig. 2i)	1	1	1	1
Capsule width (Fig. 2j)	1*	9	1*	1

O_a, states in the commonest arrangement with data were presented as bars showing 95% confidence intervals on the mean, ordinate linear; O_b, ditto, the commonest arrangement with data as 95% confidence intervals, ordinate log; O_c, ditto, the commonest arrangement with data as SD $\times$ 2 bars, ordinate linear; O_d, ditto, the overall most common arrangement; numbers are those of the arrangements reading from the top of each diagram (see Fig. 4.3). For additional information on the characters, see Gift and Stevens (1997) and Gift, Kron and Stevens (in preparation).
*There were ties for the commonest arrangement; the first arrangement in each column was chosen.

Table 4.4. Characters with non-overlapping variation used in analyses

1. Prophylls: perulate, basal (0), subfoliaceous, internode short (1), foliaceous, internode normal length (2) - N_a, N_c.
2. Phyllotaxy: leaves spiral (0), opposite (1), whorled (2) - N_b, N_c.
3. Leaves in bud: revolute (0), convolute (1) - N_a, N_c.
4. Mucilaginous epidermis in lamina: absent (0), present (1) - N_a, N_c.
5. Pith type: heterogeneous (0), homogeneous (1) - N_b, N_c.
6. Inflorescence position: terminal (0), axillary (1) - N_a, N_c.
7. Pedicel with multicellular hairs: yes (0), no (1) - N_b, N_c.
8. Calyx with midrib visible when dry: no (0), yes (1) - N_a, N_c.
9. Stomata on adaxial surface of calyx: no (0), yes (1) - N_b, N_c.
10. Stomata on abaxial surface of calyx: no (0), yes (1) - N_a, N_c.
11. Corolla pouches: absent (0), present (1) - N_b, N_c.
12. Hairs on filaments: absent (0), present (1) - N_a, N_c.
13. Anther dehiscence: by pores or short slit up to ⅓ the length of the anthers, by slits $\pm$ the length of the anthers (1) - N_b, N_c.
14. Viscin threads in pollen: present (0), absent (1) - N_a, N_c.
15. Carpel number: 5 (0), 4 (1), 3 (2) - N_b, N_c.
16. Ovary with multicellular hairs: yes (0), no (1) - N_a, N_c.
17. Style impressed in apex of ovary: yes (0), no (0) - N_b, N_c.
18. Fruit dehiscence: septicidal only (0), septicidal + loculicidal (1) - N_a, N_b, N_c.
19. Pedicel pendulous in fruit: no (0), yes (1) - N_b, N_c.

N_a, characters used in the N_a matrix; N_b, characters used in the N_b matrix; N_c, characters used in the N_c matrix. For further discussion of the characters, see Gift, Kron and Stevens (in preparation).

character states in the variation of 10 characters with overlapping variation (Gift and Stevens, 1997). Four data matrices were based on the commonest arrangements (the commonest way in which the graphs were divided into character states) recognized in this study: when data were presented as bars showing 95%

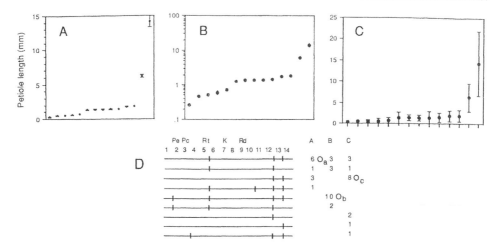

Figure 4.3 Relationship between observations (data-II) and character states (data-III) for the character, petiole length. Abscissa, the 14 taxa examined arranged in sequence of ascending measurements. A, variation represented as 95% confidence intervals on the mean. B, the same, but the ordinate $\log_{10}$. C, variation represented as bars showing $2 \times$ SD. D, division of measurements into character states; numbers along the top, taxa, with outgroup taxa indicated (Pe and Pc, *Phyllodoce empetriformis* and *P. coerulea*; Rt, *Rhodothamnus chamaecistus*, K, *Kalmiopsis leachiana*, Rd, *Rhododendron lapponicum*); bars on horizontal lines, places on graphs where subjects indicated limits of character states; column A, summary of scorings for treatment A, with O_a the state delimitation included in the O_a matrix; column B, summary of scorings for treatment B, with O_b the state delimitation included in the O_b matrix; column C, summary of scorings for treatment C, with O_c the state delimitation included in the O_c matrix. Note that the numbers in columns A, B and C differ because different subjects were involved. For further details, see Gift and Stevens (1997).

confidence intervals on the mean, the ordinate being linear (O_a); data similar, but the ordinate log-scaled (O_b); data presented as SD x 2 bars, ordinate linear (O_c); and the overall most common arrangement (O_d – Table 4.3).

The initial non-overlapping data-III set was rather larger, consisting of 23 characters and 21 taxa. Three matrices were based on this data set, each including the same taxa as in the O matrices. N_a and N_b each included ten characters from this initial matrix; since four uninformative characters were excluded, one character, 18, was arbitrarily included in both matrices, the others were different. At 10 characters, these two matrices were of equal size to the O matrices, although all the O matrices had 2½–3 times the number of states that the N_a and N_b matrices; in the latter the characters were mostly two-state. The matrix N_c included all characters in the N_a and N_b matrices (Table 4.4).

Combined (C) matrices were formed by adding the four O matrices to the N_c matrix, so forming four matrices, C_a, C_b, C_c, and C_d, each with 29 characters.

Characters were analysed as both unordered and ordered using PAUP version 4.0.0.56. A branch and bound search was carried out, with the MULPARS option and FURTHEST addition sequence. Strict and Adams consensus cladograms were calculated. Branches with >50% bootstrap support were identified; the Tree

Bisection Reconnection option was exercised, MAXTREES was set at 400, and 250 replicates were analysed. Identifying branches with >50% support focuses on branches with substantial support, since although bootstrapping yields biased values of support, the bias is fortunately usually conservative (Hillis and Bull, 1993).

Results

Groupings recognised in the various analyses are summarized in Table 4.5, and some relevant statistics in Table 4.6. *Loiseleuria + Leiophyllum* in particular, but also *Kalmia microphylla + K. polifolia*, and *K. ericoides + K. hirsuta* are strongly associated species pairs. The only other common association was *Rhodothamnus + Kalmiopsis*, although here bootstrap values were always low.

When character states are ordered, analysis of N_c, the matrix with a full complement of characters showing non-overlapping variation, alone showed an association between *Loiseleuria* and *Leiophyllum* and two species of *Kalmia*, *K. microphylla* and *K. polifolia*, but with a rather low bootstrap value. On the other hand, analysis of the O_b matrix led to an association of several species of *Kalmia*, including the two mentioned, with *Kalmiopsis*. There were several interesting associations of taxa just below the 50% bootstrap cut-off value. These included that between *Loiseleuria*, *Leiophyllum*, and the two species of *Kalmia* mentioned above at the 36% level in the C_c matrix, although the same matrix yielded an association between *Rhododendron lapponicum*, *K. ericoides* and *K. hirsuta* at only just below the 50% cut-off.

Adding the O matrices to the N_c matrix did not change groupings that had over 75% bootstrap support. However, some of the groupings evident in the analysis of the O_b matrix disappeared. When the character states were analysed as unordered, there were slight reductions in bootstrap values, and both the novel association of taxa in O_b and the association of *Leiophyllum*, *Loiseleuria*, *Kalmia polifolia* and *K. microphylla* in N_c disappeared below the 50% level. In the latter case the association was recognised at the 43% level – but so was an association between *Rhododendron lapponicum*, *K. hirsuta*, and *K. ericoides*, while *K. angustifolia* and *K. cuneata* linked at the 49.5% level.

Discussion

Since the matrices are taken from rather different kinds of characters, how can the relationships suggested be evaluated? Comparison with those evident in analysis of yet other kinds of data is one way. Unfortunately, there is no molecular phylogeny for exactly the same taxa as studied here, but Kron and King (1996) analysed variation in the ribosomal ITS + 5.8s region in seven species of *Kalmia*, *Loiseleuria*, *Leiophyllum*, and several outgroups, and in the chloroplast *rbc*L gene in three species of *Kalmia*, *Loiseleuria*, *Leiophyllum*, and outgroups. The topologies of their analyses differ in detail, but *Kalmia + Loiseleuria + Leiophyllum* form a monophyletic group, *Kalmia* being paraphyletic. In one ITS analysis *Loiseleuria + Leiophyllum* were associated with *K. polifolia* and its relatives, although the bootstrap value was only 52% (Fig. 4.4). In the combined analysis of both sequences

Table 4.5. Groupings recognized in the analyses

Matrices	Lois Leio	K. micro K. polif	Lois Leio K. micro K. polif	K. eric K. hirs	R'tham K'opsis	P. coer P. empet	K. micro K. poli K. cune K. latif K. angus K'opsis	K. polif K. cune K. latif K. angus	K. cune K. latif K. angus	K. latif K. angus
N_a	*			*	*					
N_b	**									
N_c	**	*		**	**					
O_a	**									
O_b	**			*		*	*	*	*	*[1]
O_c	**			*			*			
O_d	**			**	*					
C_a	**			**	*	*				
C_b	**			**	*	*			**	
C_c	**			**	*					*
C_d	**			**						

For full names, see Table 4.2.
** Bootstrap value >75%, * bootstrap value 50-75%.
[1] This grouping appeared in only one bootstrap run.

Table 4.6 Summary statistics of the analyses

	Non-overlapping			Overlapping				Combined			
	N_a	N_b	N_c	O_a	O_b	O_c	O_d	C_a	C_b	C_c	C_d
Character/states	10/11	10/11	20/23	10/32	10/29	10/24	10/25	30/55	30/52	30/47	30/48
Consistency index	.53	.60	.54	.53	.67	.58	.60	.50	.56	.51	.53
Homoplasy index	.48	.40	.46	.47	.33	.42	.40	.50	.44	.49	.47
Retention index	.72	.73	.68	.56	.72	.63	.64	.57	.63	.59	.61
Rescaled consistency index	.38	.44	.38	.30	.48	.39	.38	.29	.38	.31	.33

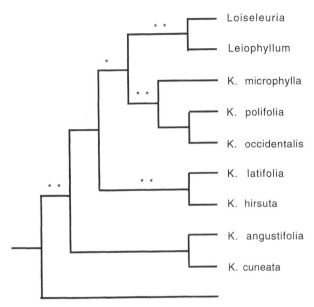

Figure 4.4 Relationships within *Kalmia* s.l (from Kron and King, 1996). Single asterisks represent bootstrap values of 50–75%; double asterisks, values over 75%.

the grouping (((*Loiseleuria* + *Leiophyllum*) *K. latifolia*) *K. polifolia*) was strongly supported.

If it is assumed that the molecular phylogenies reflect true phylogenetic relationships, morphology in the analyses described here does rather poorly, hinting at an association between *Kalmia* and *Loiseleuria* in only one analysis, that of the full morphological data matrix with ordered states. In general, the O matrices performed still more poorly than the N matrices, but there is the hint of an entirely novel set of relationships in the analysis of the O_b matrix. Here variation was displayed as 95% confidence intervals on the mean with a $\log_{10}$ ordinate – which would seem not to be common practice among those using such measurement data. In any case, the association of *Kalmia polifolia*, some other species of *Kalmia*, and *Kalmiopsis* is weakly supported, not found in other analyses, and unlikely to reflect genealogical relationships, i.e. it positively misleads.

I was unable to recapture such relationships in analyses of the N matrices. On the other hand, a monophyletic *Kalmia* almost never appears unless the number of outgroups is severely reduced, e.g. to *Phyllodoce* and *Rhodothamnus*. Many species groupings, including ((*K. polifolia* + *K. microphylla* + *K. glauca*) (*Loiseleuria* + *Leiophyllum*)), appear if the outgroups are reduced to *Epigaea repens*, *Kalmiopsis leachiana* and *Rhodothamnus chamaecistus* (see Gift, Kron and Stevens, in prep., for details).

As Gift and Stevens (1997) showed, different ways of presenting data-II affect how character states in those data are delimited, hence the four O matrices used here are alternative ways of dividing the 'same' data-II. For all characters, the overall most common arrangement (O_d) was always one of the commonest arrangements in an individual treatment ($O_{a, b, c}$). This may not always be the case. An arrangement can be recognised by over half the subjects, yet not be any of the arrangements most common in one treatment. This nearly happened for two characters, emphasizing how complex delimiting character states in such data is likely to be.

General discussion

The results of single studies do not 'prove' anything, the more so since the analyses above are a small subset of those possible. However, other evidence suggests that states based on overlapping variation are not as effective in uncovering genealogies as are states based on non-overlapping variation (Stevens, 1991; Gift and Stevens, 1997, for references), although this may not always be the case (G. Levin, personal communication). Clearly, this study does nothing to bolster confidence in the use of such variation. Extreme caution is in order when interpreting the topologies of cladograms which are based on an appreciable number of characters in which variation is of the O-type, and for which states were determined by visual inspection of data-I or data-II.

Statistical techniques may allow O-type variation to be utilized in phylogenetic studies (e.g. Strait *et al.*, 1996; Rae, 1998). But as with visual inspection of such variation, we know little about how such techniques perform. Thiele (1994) analysed matrices derived from O and N ('qualitative') variation in three sections of *Banksia*, and found considerable topological similarities between the cladograms produced in separate analyses of the data-III. Unfortunately, there was no molecular phylogeny with which to compare these results, nor were bootstrap values given.

The distinction between characters in which the variation is overlapping and that in which it is non-overlapping is of course not itself clearcut. As Thiele (1994; see also Kitching *et al.*, 1998) noted, variation in many characters can be described in terms of frequency distributions. This is particularly true of variation in carpel number, here somewhat inaccurately described as being non-overlapping. *Loiseleuria* has two carpels, only rarely three; about two-thirds of the flowers of *Leiophyllum* are three-carpellate, the rest have two carpels. Scoring the two taxa as having different states is equivalent to saying that the frequencies of two- and three-carpellate gynoecia in the two taxa are very different (see Thiele, 1994). Similarly, rare meristic variants occur in *Kalmia*, but little is gained by treating such variation as being problematical. The N and O data matrices are quantitatively different in the degree to which the states show overlap, even if they are not qualitatively different.

If we are interested in understanding more about how measurement and observation convert into character states, one road ahead is clear. Studies of groups of organisms in which different kinds of data are produced for each taxon, and then analysed separately and in combination, are of the highest priority. Comparing topologies and their support will clarify just what aspects of the organisms we observe are of most use in uncovering phylogenies. Of course, we have to guard against ideas of relationship gained using one kind of data affecting the delimitation of states when using another kind (Stevens, 1991; Gift and Stevens, 1997).

If O-type data-I and data-II are used, the adoption of explicit methods to determine the existence of gaps may add a step to cladistic studies. Nevertheless, the actual presentation of data-II prior to deciding on the existence of character states can be quite simple (Gift and Stevens, 1997; Swiderski et al., 1998). If characters showing overlapping variation are excluded from analyses, there may be only a few characters left. However, the issue is quality versus quantity; cladograms produced from questionable data will not engender confidence. Indeed, Pennington (1996) discarded most morphological characters in his phylogenetic study of *Andira* because of the problem of overlapping variation, and only ten characters remained. Yet he found these few characters useful in clarifying structure at low hierarchical levels; the restriction site data he also used was more useful at detecting major structure.

Unsubstantiated assertions that characters used in a particular study show non-overlapping variation simply will not do. Visual analysis of overlapping variation, whether summarized and presented as graphs or dependent on comparisons of spec-imens or descriptions, is a complex and little-understood process with few guidelines in the literature. There is no single way in which all observers see pattern in such variation; indeed, what is considered to be pattern will depend on details of an individual's education and even general upbringing. Yet O-type characters are commonly used in systematic studies (Stevens, 1991; Gift and Stevens, 1997); furthermore, many characters presented as if the underlying variation were quali-tative in fact represent semantically transformed quantitative variation (Stevens, 1991; cf. also Stuessy, 1990).

Auditory or spectral sense data present similar problems in terms of perception. Organisms as diverse as cladists and crickets may vehemently assert that there are categorically different states in the stimuli, even though those underlying stimuli are part of a continuum. We can distinguish accurately between small differences in the wavelengths of two light stimuli, yet we may label these small differences as being categorically distinct, or we may even perceive differences in sound frequen-cies categorically – and of course we then label them categorically as well (Wyttenbach et al., 1996). For some behavioural studies we may need to use the distinctions made by other organisms; for others, whatever pattern is evident on visual inspection or statistical analysis of curves of frequency distributions.

The issue of the delimitation of character states or their equivalent is critical for biological comparisons in general. If states are not obvious, or can be delim-ited differently, or clusters in a graph are not obvious, then evolutionary scenarios based on such studies are questionable (e.g. Dale, 1995; Garwood, 1995; Manly, 1996; Waser et al., 1996; Wiens and Morris, 1996; Solow and Beet, 1998). Does it make sense to treat morphological and molecular data as undivided categories when comparing them? Understanding more about the kinds of decisions involved in producing phylogenetic data provides a different perspective to such problems.

It can be argued that how states are delimited in morphological studies matters less than we might think, since it is unlikely that any systematic error is introduced into phylogenetic analyses (Donoghue, personal communication). But then it is diffi-cult to see what might be gained by making comparisons between different kinds of data; it is very easy to add characters showing overlapping size variation to morphological analyses, yet one is adding (a) characters that may be divided into states differently by different people and (b) data that in some cases may simply be noise. Along these lines, Nixon and Carpenter (1996: 223) suggested that 'total evidence' could include continuous quantitative (= non-overlapping) data that might be useful 'in ways other than a parsimony analysis' (cf. also Meacham, 1994). Although Donoghue and Sanderson (1992: 359) professed themselves 'suspicious of assertions that some forms of data are useless', we have to go beyond both suspicions and assertions. And as one moves from trees to phylogenies, to talk about evolution in characters such as those that are the focus of this paper makes little sense. Although I agree with Kitching *et al.* (1998) that quantitative, qualitative, continuous and discrete do not refer to different kinds of data, dividing data into admittedly imprecise categories can clarify under what circumstances we can put confidence in their use, just as distinguishing night from day can on occasion be helpful.

I conclude by quoting from a talk given by William T. Stearn at a Systematics Association meeting 35 years ago: 'both [the taxonomist and artist] seek pattern within diversity, the one to record those he thinks he finds in nature, the other to record those he finds maybe only in his own head' (Stearn, 1964: 84). It has been common to think that a systematist's business was 'observing nature', and that this was a largely unproblematical operation, especially when guided by sound taxonomic instinct (Stevens, 1994, 1998). J.S.L. Gilmour, a key figure in the founding of the Systematics Association (Winsor, 1995), confirmed this general view. In a very influential paper, 'Taxonomy and philosophy' (Gilmour, 1940), he distinguished between sense data, which he considered were given once and for all and unalterable, and the clips, the different ways in which these sense data could be analysed to form classifications. The advent of phenetics and cladistics has not challenged this approach. The patterns that Stearn was discussing, admittedly perhaps a little tongue-in-cheek, are not self-evident; nature is not unambiguous. Recent work suggests that Stearn's taxonomist has been functioning more like how he saw artists functioning. Many plant groups, recognised as such in part because systematists had always recognised them, have distressingly less to do with phylogeny than might have been hoped. Even at the level of delimitation of states we find that there are often no simple patterns in nature that force themselves on our senses.

Summary

A comparison of how we decide that character states can be abstracted from basic molecular and morphological observations emphasizes how difficult it can be to delimit states in morphological data. However, the initial decisions that identify molecular and morphological characters are rather similar if we allow for the greater body of morphological knowledge. This helps to clarify the discussion surrounding the nature of the two kinds of data. In studies of two kinds of morphological data, one in which the states are taken from visual inspection of overlapping variation and one in which states are taken from largely non-overlapping variation, the latter was found to contain more phylogenetic signal. Here, too, recognising that not all data are necessarily of equal value allows us to understand the strengths and limitations of what we are trying to do, even if it cannot allow us to reject categorically particular kinds of data as being of no help in our work.

Acknowledgements

I am very grateful to M.J. Donoghue, A. Doust, L. Hufford, E.A. Kellogg, G.A. Levin, R. Mason-Gamer, R. Olmstead, R.T. Pennington, R.W. Scotland, P. Weston, the systematics discussion group at the Harvard University Herbaria and an anonymous reviewer for variously reading all or parts of the manuscript, advice and discussion.

References

Bock, W.J. (1974) Philosophical foundations of classical evolutionary classification, *Systematic Zoology*, **22**, 375–392.

Bock, W.J. (1977) Foundations and methods of evolutionary classification, in Hecht, M.K., Goody, P.C. and Hecht, B.M. (eds) *Major Patterns in Vertebrate Evolution*, New York: Plenum, pp. 851–895.

Boyden, A. (1943) Homology and analogy: a century after the definitions of 'homologue' and 'analogue' of Richard Owen, *Quarterly Review of Biology*, 18, 228–241.

Brady, R.H. (1994) Explanation, description, and the meaning of 'transformation' in taxonomic evidence, in Scotland, R.W., Siebert, D.J. and Williams, D.M. (eds) *Models in Phylogenetic Reconstruction*, Oxford: Clarendon Press, pp. 11–29.

Brower, A.V.Z. and Schawaroch, V. (1996) Three steps of homology assessment, *Cladistics*, **12**, 265–272.

Burtt, B.L. (1997) Old World Gesneriaceae V. suprageneric names, *Edinburgh Journal of Botany*, **54**, 85–90.

Colless, D. (1985) On "character" and related terms, *Systematic Zoology*, **34**, 229–233.

Crowe, T.M. (1994) Morphometrics, phylogenetic models and cladistics: means to an end or much to do about nothing? *Cladistics*, **10**, 77–84.

Dale, M.B. (1995) Evaluating classification strategies, *Journal of Vegetation Science*, **6**, 437–440.

Davison, D. (1985) Sequence similarity ('homology') searching for molecular biologists, *Bulletin of Mathematical Biology*, **47**, 437–474.

De Pinna, M.C.C. (1991) Concepts and tests of homology in the cladistic paradigm, *Cladistics*, **7**, 367–394.

Donoghue, M.J. (1992) Homology, in Keller, E.F. and Lloyd, E.A. (eds) *Keywords in Evolutionary Biology*, Cambridge, MA: Harvard University Press, pp. 170–179.

Donoghue, M.J. and Sanderson, M.J. (1992) The suitability of morphological and molecular evidence in reconstructing plant phylogeny, in Soltis, P.S., Soltis, D.E. and Doyle, J.J. (eds) *Molecular Systematics of Plants*, New York: Chapman and Hall, pp. 340–368.

Donoghue, M.J. and Sanderson, M.J. (1994) Complexity and homology in plants, in Hall, B.K. (ed.) *Homology: the Hierarchical Basis of Comparative Biology*, San Diego: Academic Press, pp. 393–421.

Doyle, J.J. (1992) Gene trees and species trees: molecular systematics as one-character taxonomy, *Systematic Botany*, **17**, 144–163.

Doyle, J.J. (1997) Trees within trees: genes and species, molecules and morphology, *Systematic Biology*, **46**, 537–553.

Eldredge, N. (1979) Alternative approaches to evolutionary theory, *Bulletin of the Carnegie Museum of Natural History*, **13**, 7–19.

Farris, J.S. (1990) Phenetics in camouflage, *Cladistics*, **6**, 91–100.

Fink, W.L. and Zelditch, M.L. (1995) Phylogenetic analysis of ontogenetic shape transformations: a reassessment of the piranha genus *Pygocentrus* (Teleostei), *Systematic Biology*, **44**, 343–360.

Fisher, D.R. and Rohlf, F.J. (1969) Robustness of numerical taxonomic methods and errors in homology, *Systematic Zoology*, **18**, 33–36.

Fitter, A.H. (1995) Interpreting quantitative and qualitative data in comparative analyses, *Journal of Ecology*, **83**, 730.

Fristrup, K. (1992) Character: current usages, in Keller, E.F. and Lloyd, E.A. (eds), *Keywords in Evolutionary Biology*, Cambridge, MA: Harvard University Press, pp. 45–51.

Garwood, N.C. (1995). Studies in Annonaceae – XX. Morphology and ecology of seedlings, fruits and seeds of selected Panamanian species, *Botanischer Jahrbücher*, **117**, 1–152.

Gift, N. and Stevens, P.F. (1997) Vagaries in the delimitation of character states in quantitative variation – an experimental study, *Systematic Biology*, **46**, 112–125.

Gilmour, J.S.L. (1940) Taxonomy and philosophy, in Huxley, J. (ed) *The New Systematics*, Oxford: Oxford University Press, pp. 461–474.

Goodman, N. (1972) Seven strictures on similarity, in Goodman, N. (ed.) *Problems and Projects*, New York: Bobbs-Merrill, pp. 437–447.

Hall, B.K. (ed.)(1994) *Homology: the Hierarchical Base of Comparative Biology*, San Diego, CA: Academic Press.

Hawkins, J.A., Hughes, C.E. and Scotland, R.W. (1997) Primary homology assessment, characters and character states, *Cladistics*, **13**, 275–283.

Hay, A. and Mabberley, D.J. (1994) On perception of plant morphology: some implications for phylogeny, in Ingram, D.S. and Hudson, A. (eds) *Shape and Form in Plants and Fungi*, London: Academic Press, pp. 101–117.

Hillis, D.M. and Bull, J.J. (1993) An empirical test of bootstrapping as a method for assessing confidence in phylogenetic analysis, *Systematic Biology*, **42**, 182–192.

Hillis, D.M., Mable, B.K., Larson, A., Davis, S.K. and Zimmer, E.A. (1996) Nucleic acids IV: sequencing and cloning, in Hillis, D.M., Moritz, C. and Mable, B.K. (eds) *Molecular Systematics*, 2nd edn, Sunderland, MA: Sinauer, pp. 321–381.

Inglis, W.C. (1966). The observational basis of homology, *Systematic Zoology*, **15**, 219–228.

Jardine, N. (1967) The concept of homology in biology, *British Journal for the Philosophy of Science*, **18**, 125–139.

Jardine, N. (1969) The observational and theoretical components of homology: a study on the morphology of dermal skull-roofs of riphidistrian fishes, *Biological Journal of the Linnean Society*, **1**, 327–361.

Kellogg, E.A. and Juliano, N.D. (1997) The structure and function of RuBisCO and their implications for systematic studies, *American Journal of Botany*, **84**, 413–428.

Kesner, M.H. (1994) The impact of morphological variants on a cladistic hypothesis with an example from a myological data set, *Systematic Biology*, **43**, 41–57.

Kitching, I.J., Forey, P.L., Humphries, C.J., and Williams, D.M. (1998) *Cladistics: the Theory and Practice of Parsimony Analysis*, 2nd edn, Oxford: Oxford University Press.

Kron, K.A. and King, J.M. (1996) Cladistic relationships of *Kalmia*, *Leiophyllum*, and *Loiseleuria* (Phyllodoceae, Ericaceae) based on *rbc*L and nrITS data. *Systematic Botany*, **21**, 17–29.

McLellan, T. and Endler, J.A. (1998) The relative success of some methods for measuring and describing the shape of complex objects, *Systematic Biology*, **47**, 264–281.

McShea, D.W. (1996) Complexity and homoplasy, in Sanderson, M.J. and Hufford, L. (eds) *Homoplasy: The Recurrence of Similarity in Evolution*, San Diego, CA: Academic Press, pp. 207–225.

Manly, D.F.J. (1996) Are there clumps in body size distributions? *Ecology*, **77**, 81–86.

Meacham, C.A. (1994) Phylogenetic relationships at the basal radiation of the angiosperms, *Systematic Botany*, **19**, 506–522.

Medin, D.L., Goldstone R.L. and Gentner, D. (1993) Respects for similarity, *Psychological Review*, **100**, 254–278.

Meeuse, A.D.J. (1975) Changing floral concepts: Anthocorms: flowers and anthoids, *Acta Botanica Neerlandica*, **24**, 23–36.

Meeuse, A.D.J. (1984) Homology as an empiricism, *Journal of Plant Anatomy and Morphology*, **1**, 9–24.

Melchior, H. (1930) *Decaphalangium*, eine neue Gattung der Guttiferen aus Peru, *Notizblatt Botanischer Garten Berlin*, **10**, 946–950.

Melville, R. (1983) Remoration: an overlooked concept in angiosperm evolution, *Kew Bulletin*, **37**, 613–632.

Mindell, D.P. (1991a) Aligning DNA sequences: homology and phylogenetic weighting, in Miyamoto, M.M. and Cracraft, J. (eds) *Phylogenetic Analyses and DNA Sequences*, New York: Oxford University Press, pp. 73–89.

Mindell, D.P. (1991b) Similarity and congruence as criteria for molecular homology, *Molecular Biology and Evolution*, **8**, 897–900.

Miyamoto, M.M. and Cracraft, J. (1991) Phylogenetic inference, DNA sequence analysis, and the future of molecular systematics, in Miyamoto, M.M. and Cracraft, J. (eds) *Phylogenetic Analyses and DNA Sequences*, New York: Oxford University Press, pp. 3–17.

Moritz, C. and Hillis, D.M. (1996) Molecular systematics: context and controversies, in Hillis, D.M., Moritz, C. and Mable, B.K. (eds) *Molecular Systematics*, 2nd edn, Sunderland, MA: Sinauer, pp. 1–3.

Nixon, K.C. and Carpenter, J.M. (1996) On simultaneous analysis, *Cladistics*, **12**, 221–241.

Norman, G.R., Brooks, L.R., Coblentz, C.L. and Babcock, C.J. (1992) The correlation of feature identification and category judgements in diagnostic radiology, *Memory and Cognition*, **20**, 344–355.

Panchen, A.L. (1994) Richard Owen and the concept of homology, in Hall, B.K. (ed.) *Homology: The Hierarchical Basis of Comparative Biology*, San Diego: CA: Academic Press, pp. 21–62.

Patterson, C. (1982) Morphological characters and homology, in Joysey, K.A. and Friday, A.E. (eds) *Problems in Phylogenetic Reconstruction*, London: Academic Press, pp. 21–74.

Patterson, C. (1988) Homology in classical and molecular biology, *Molecular Biology and Evolution*, **5**, 603–625.

Patterson, C. and Johnson, C.D. (1997) The data, the matrix, and the message: comments on Begle's "Relationships of the osmeroid fishes", *Systematic Biology*, **46**, 358–365.

Pennington, R.T. (1996) Molecular and morphological data provide phylogenetic resolution at different hierarchical levels in *Andira*, *Systematic Biology*, **49**, 416–515.

Pimentel, R.A. and Riggins, R. (1987) The nature of cladistic data, *Cladistics*, **3**, 201–209.

Platnick, N.I. (1979) Philosophy and the transformation of cladistics, *Systematic Zoology*, **28**, 537–546.

Pleijel, F. (1995) On character coding for phylogeny reconstruction, *Cladistics*, **11**, 309–315.

Rae, T.C. (1998) The logical basis for the use of continuous characters in phylogenetic systematics, *Cladistics*, **14**, 221–228.

Raikow, R.J., Bledsoe, A.H., Myers, B.A. and Welsh, C.J. (1990) Individual variation in avian muscles and its significance for the reconstruction of phylogeny, *Systematic Zoology*, **39**, 362–370.

Remane, A. (1952) *Die Grundlagen des natürlichen Systems, der vergleichenden Anatomie und der Phylogenetick*, Leipzig: Akademische Verlagsgesellschaft.

Riedl, R. (1976) *Die Strategie der Genesis*, Zürich: R. Piper.

Rieppel, O.C. (1988) *Fundamentals of Comparative Biology*, Basel: Birkhäuser.

Rieppel, O.C. (1994) Homology, topology, and typology: The history of modern debates, in Hall, B.K. (ed.) *Homology: The Hierarchical Basis of Comparative Biology*, San Diego, CA: Academic Press, pp. 63–100.

Sanderson, M.J. and Donoghue, M.J. (1996) The relationship between homoplasy and confidence in a phylogenetic tree, in Sanderson, M.J. and Hufford, L. (eds) *Homoplasy: the Recurrence of Similarity in Evolution*, San Diego, CA: Academic Press, pp. 67–89.

Sattler, R. (1992) Process morphology: Structural dynamics in development and evolution. *Canadian Journal of Botany*, 70, 708–716.

Sattler, R. and Rutishauser, R. (1990) Structural and dynamic descriptions of the development of *Utricularia foliosa* and *U. australis*, *Canadian Journal of Botany*, 68, 1989–2003.

Smith, A. (1994) *Systematics and the Fossil Record*, Oxford: Blackwell Scientific.

Sneath, P.H.A. and Sokal, R.R. (1973) *Numerical Taxonomy*, San Francisco, CA: W.H. Freeman.

Sokal, R.R. and Sneath, P.H.A. (1963) *The Principles of Numerical Taxonomy*, San Francisco, CA: W.H. Freeman.

Solow, A.R. and Beet, A.R. (1998) On lumping species in food webs. *Ecology* 79, 2013–2018.

Stearn, W.T. (1964) III. Problems of character selection and weighting. Introduction, in Heywood, V.H., and McNeill, J. (eds) *Phenetic and Phylogenetic Classification*, London: Systematics Association, pp. 83–86.

Stevens, P.F. (1984) Homology and phylogeny: morphology and systematics. *Systematic Botany* 9, 395–409.

Stevens, P.F. (1991) Character states, morphological variation, and phylogenetic analysis: a review. *Systematic Botany* 16: 553–583.

Stevens, P.F. (1994) *The Development of Biological Systematics*, New York: Columbia University Press.

Stevens, P.F. (1996) On phylogenies and data bases – where are the data, or are there any? *Taxon* 45, 95–98.

Stevens, P.F. (1998) Mind, memory, and history: how classifications are shaped by and through time, and some consequences, *Zoologica Scripta* 26, 293–301.

Stevens, P.F. (1999) On the relationships of *Decaphalangium* and *Renggeria* and an enumeration of species in *Clusia* section *Cordylandra* (Clusiaceae) with comments on androecial morphology, *Sida*, accepted for publication.

Strait, D.S., Moniz, M.A. and Strait, P.T. (1996) Finite mixture coding: A new approach to coding continuous characters, *Systematic Biology* 45, 67–78.

Stuessy, T.F. (1990) *Plant Taxonomy: The Systematic Evaluation of Comparative Data*, New York: Columbia University Press.

Swiderski, D.L., Zelditch, M.L. and Fink, W.L. (1998) Why morphometrics isn't special: coding quantitative data for phylogenetic analysis. *Systematic Biology*, 47, 508–519.

Thiele, K. (1994) The holy grail of the perfect character: The cladistic treatment of morphometric data. *Cladistics* 9: 275–304.

Thom, R. (1992) Pouvoirs de la forme, in Gayon, J. and Wunenburger, J.-J. (eds), *Les Figures de la Forme*, Paris: L'Harmattan, pp. 17–26.

Wagner, G.P. (1989) The biological homology concept, *Annual Review of Ecology and Systematics*, 20, 51–69.

Wagner, G.P. (1994) Homology and the mechanisms of development, in Hall, B.K. (ed.), *Homology: the Hierarchical Basis of Comparative Biology*, San Diego, CA: Academic Press, pp. 273–299.

Wake, M.H. (1994) The use of unconventional morphological characters in the analysis of systematic patterns and evolutionary process, in Grande, L. and Rieppel, O. (eds), *Interpreting the Hierarchy of Nature*, San Diego, CA: Academic Press, pp. 173–200.

Walters, S.M. (1964) General Discussion. Introduction, in Heywood, V.H. and McNeill, J. (eds), *Phenetic and Phylogenetic Classification*, London: Systematics Association, pp. 157–159.

Waser, N.M., Chittka, L., Price, M.V., Williams, N.M. and Ollerton, J. (1996). Generalization in pollination systems, and why it matters. *Ecology* 77, 1043–1060.

Waterman, M.S., Joyce J. and Eggert, M. (1991) Computer alignment of sequences, in Miyamoto, M.M., and Cracraft, J. (eds) *Phylogenetic Analyses and DNA Sequences*, New York: Oxford University Press, pp. 59–72.

Wheeler, W.C. (1995) Sequence alignment, parameter sensitivity, and the phylogenetic analysis of molecular data, *Systematic Biology* 44, 321–331.

Wheeler, W.C. (1996) Optimization alignment: the end of multiple sequence alignment in phylogenetics? *Cladistics* 12, 1–9.

Wheeler, W.C. and Gladstein, D. (1992) *MALIGN*. New York: American Musuem of Natural History.

Wiens, J.J. and Morris, M.R. (1996) Character definitions, sexual selection, and the evolution of swordtails. *American Naturalist* 147, 866–869.

Wilkinson, M. (1995) A comparison of two methods of character construction. *Cladistics* 11, 297–308.

Winsor, M.P. (1995) The English debate on taxonomy and phylogeny, 1937–1940. *History and Philosophy of the Life Sciences* 17, 227–252.

Woodger, J.H. (1945) On biological tranformations, in Le Gros Clark, W.E. and Medawar, P.B. (eds), *Essays on Growth and Form presented to D'Arcy Wentworth Thompson*, Oxford: Clarendon Press, pp. 95–120.

Wyttenbach, R.A., May, M.L. and Hoy, R.R. (1996) Categorical pereception of sound frequency by crickets. *Science* 273, 1542–1544.

Zelditch, M.L., Fink, W.L. and Swiderski, D.L. (1995) Morphometrics, homology, and phylogenetics: quantified characters as synapomorphies, *Systematic Biology* 44, 179–189.

Zimmermann, W. (1965) *Die Telomtheorie*, Stuttgart: Gustav Fischer.

Chapter 5

Heuristic reconstruction of hypothetical–ancestral DNA sequences: sequence alignment vs direct optimization

Ward Wheeler

Introduction

The problem of historical reconstruction of nucleic acid sequences can be reduced to one of the determination of ancestral (i.e. nodal) sequences. The composition of these hypothetical sequences determines the length and shape of any cladogram. The efficient or parsimonious placement of these nodes is the fundamental systematic act.

Nucleic acid (and protein) sequence data present problems not normally seen in other types of comparative data. The main distinction is that the terminal taxa do not present the same number of features. This, coupled with the limited number of character states, causes the correspondences among features to be undetermined. In other words, the putative homologies among the sequence bases are unknown. Most anatomical data do not have this problem. A feature of a vertebrate fore-limb will never be compared to the retinular structure of the eye or even a superficially similar hind limb. The positional information implicit in the character definition and the effectively infinite number of states leaves no room to confuse character states among characters (the case described by Whiting and Wheeler (1994) is presumed to be rare).

The normal course of events for the systematic analysis of sequence information begins with some form of multiple sequence alignment. Although some workers still argue for alignments 'by eye', this technique clearly is not based on any quantitative optimality criterion, is highly subject to preconceived notions of relationship, and is non-reproducible. Algorithmically (i.e. computer-) generated multiple alignments may not be without fault, but at least they can be reproduced by other investigators. These alignments embody the putative synapomorphies on which standard phylogenetics relies. After alignment, character reconstructions may be completely isomorphic (unordered or non-additive) or specified in great detail with step matrices, and may even include asymmetrical character transformation costs.

A method has been proposed recently which seeks to optimize the sequence characters directly, without the intervening alignment step (Wheeler, 1996). The thrust of this idea is the generalization of character optimization techniques to allow for terminals with unequal numbers of characters. Existing optimization techniques such as those of Farris (1970), Fitch (1971) and Sankoff and Rousseau (1975) all require equal numbers of characters among all taxa. Once optimization has been generalized to accommodate sequence length variation, multiple sequence alignment is unnecessary.

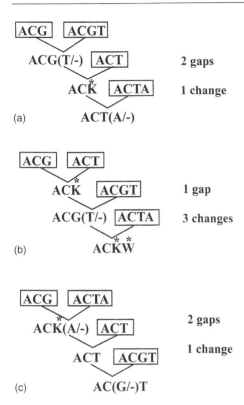

(a)

(b)

(c)

Figure 5.1 Examples of direct optimization (optimization–alignment).

In order to compare these two methods (sequence alignment and direct optimiza-tion), some measure must be agreed upon to assess their relative merits. Given the unknowability of 'truth' and the problems of simulation, congruence is the most appropriate means of comparing methods. Certainly consistency of results over multiple sources of data is a desirable property of any historical method. There may be others, but for the purpose of this discussion, consideration is limited to the congruence of systematic results in the face of diverse phylogenetic infor-mation.

Here, the methodologies of multiple sequence alignment and direct optimization are discussed and compared empirically with respect to character congruence as measured by the Mickevich and Farris (1981) (MF) incongruence length difference in three real data sets.

Direct optimization/optimization alignment

A simple restatement of the procedure described earlier (Wheeler, 1996) is shown in Fig. 5.1. In this example, four simple sequences are diagnosed on three clado-grams. As I have pointed out elsewhere (Wheeler, 1998), this algorithmic procedure is a heuristic determination of minimum tree length. The exact case is computa-tionally prohibitive (within current thinking). Since the heuristic method is greedy and short-sighted, not examining all the global possibilities, the tree length returned by this algorithm is an upper bound on the minimal length.

The essence of the method is the determination of efficient hypothetical ancestral sequences at each internal node. In short, this down-pass sequence is calculated as the sequence, which requires the minimum change to its two descendants. In practice, the sequence is calculated via dynamic programming (as with the Needleman-Wunsch (1970) alignment algorithm), with the exception that the procedure minimizes the sum cost of transformation from the nodal sequence to its descendants. The sum cost will depend on the relative gap cost and any transition:transversion bias or more complex character model. Where there are several minimal length base (or indel) assignments possible, ambiguities are noted in the nodal sequence. The length of the cladogram is determined by summing the cost of the creation of each of these nodes until the cladogram is complete. The node sequences created during the down-pass are not the final hypothetical ancestral sequences. As with standard analysis, an up-pass is required to get the final character states and branch lengths. The final nodal sequences are created to minimize distance among the three nodal sequences (final ancestor and two preliminary descendants) connected to it.

In the case of Fig. 5.1, not only optimization but also the choice of optimal topology would be affected by differential gap and change cost. If indels were relatively expensive compared to base substitutions, topology 'b' would be minimal, whereas 'a' would be favoured with relatively expensive base changes.

Data

The three data sets discussed here are derived from literature sources. They are: (1) Chelicerata – 34 taxa with 93 morphological characters, approximately 1000 bp 18S rDNA, and approximately 350 bp 28S rDNA (Wheeler and Hayashi, 1998); (2) Carnivora – 39 taxa with 265 bp mt cytochrome b, approximately 210 bp mt 12S rDNA, and approximately 310 bp mt 16S rDNA (based on Vrana et al., 1994); and (3) Hemiptera – 21 taxa with approximately 780 bp 18S rDNA, approximately 350 bp 28S rDNA, approximately 570 bp mt cytochrome oxidase I, and approximately 400 bp mt 16S rDNA (based on Wheeler et al., 1993).

Methods

Multiple sequence alignment

The program MALIGN (Wheeler and Gladstein, 1992, 1994) was used to align the sequences. In each case, the indel (gap) cost (penalty) was 2 and the base substitution cost was 1 (for both transitions and transversions). Each gap inserted in a sequence was treated independently, i.e. a gap of length three cost three times the individual gap cost. Ten random addition sequences were performed; each was then subjected to TBR branch swapping on the alignment tree. This means that many (millions in fact) multiple alignment orders were tried and for each a multiple alignment created, and a cladogram search performed. The cost of each alignment was determined by the length of the shortest tree derived from the multiple alignment.

Table 5.1 Character incongruence among data sets by method

Taxon	Data set	Direct	Malign	MF direct	MF Malign
Hemiptera	All	3263	3764	0.0362	0.0404
	16	792	850		
	18	632	781		
	28	772	1013		
	COI	949	968		
Carnivora	All	3802	4365	0.0326	0.0575
	12	853	987		
	16	1337	1666		
	Cyb	1488	1461		
Chelicerata	All	4691	5686	0.0284	0.0990
	Morph	804	804		
	18	1845	2205		
	28	1909	2114		
	Mol	3849	4539	0.0247	0.0485

Direct = direct optimization (optimization–alignment); Malign = multiple alignment produced by Malign; MF = Michevich-Farris incongruence value; COI = cytochrome oxidase I; Cyb = cytochrome B; all = combined data; 12 = 12S rDNA; 16 = 16S rDNA; 18 = 18S rDNA; 28 = 28S rDNA; mol = combined molecular data; morph = morphological data.

When these tree searches were performed SPR branch swapping was performed on multiple trees. The reason the less stringent searches were used for the choice of multiple alignment was that tree searches were performed for each one of the millions of multiple alignments performed. After the best (lowest cost) alignment was found, the program PHAST (Goloboff, 1996) was used to search for the shortest cladogram based on the best alignment using TBR branch-swapping and 10 random addition sequences.

Direct optimization

The program POY (Gladstein and Wheeler, 1997) was used to perform the direct analysis of the sequence data. As with alignment, the indel (gap) cost was 2 and the base substitution cost was 1 (for both transitions and transversions) and gap costs were linear with respect to length. Ten random addition sequences were performed; each was then subjected to TBR branch swapping. Since the results of the direct optimization of the sequences are parsimonious topologies, no further operations were required.

In both cases, the programs (POY and MALIGN) were executed on a parallel cluster consisting of 23 Unix workstations of various types united by Parallel Virtual Machine (PVM v. 3.4).

Results

The results of the combined and separate analyses are given in Table 5.1. Of the 13 analyses performed by both direct optimization and multiple alignment, 12 cases resulted in a shorter cladogram for the direct analysis than for that based on multiple alignment. This has been shown before (Wheeler, 1996) and appears to be general at

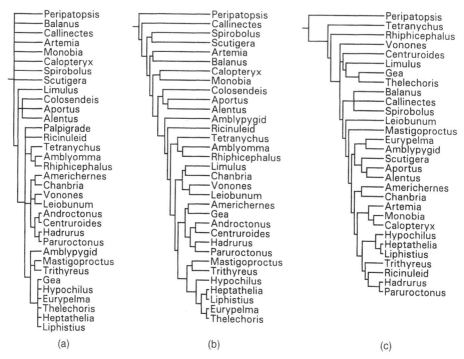

Figure 5.2 Chelicerate cladograms generated by direct optimization of separate analyses: (a) 93 morphological characters; (b) 18S rDNA; (c) 28S rDNA. The character incongruence level of 2.6% (MF) is much lower than any topological measure of congruence would suggest, since they share only four of 32 non-trivial groups.

least as far as these sequences are concerned. The enhancement of parsimony is most likely due to the fact that the putative homologies of bases in direct optimization are determined for each topology separately. In multiple alignment, these correspondences are predetermined and invariant with respect to the phylogenetic topology. The single case of alignment yielding more parsimonious results, carnivore cytochrome b, contained a great deal of missing data and may be more complex than the lengths suggest. The term 'shorter' in this case signifies that fewer insertions, deletions, and base substitutions were required by the direct analyses than those based on multiple alignment. Although I feel that these numbers are directly comparable, others have suggested that since the sequences are treated in such different ways, the simple lengths of the cladograms are not comparable. Also, MALIGN may well not be the best way to generate parsimonious multiple alignments. Assuming these factors, the most important criterion for comparison is the MF incongruence measures.

In each of the three cases examined – Hemiptera, Carnivora, and Chelicerata – the analyses based on direct optimization had lower MF incongruence levels than those derived from multiple alignment. The topological implications of these analyses are shown for the chelicerate topologies generated by direct optimization (Fig. 5.2). As can be seen in this example, although character congruence is very high, there is little common resolution among these topologies.

Data Set 'A'   Data Set 'B'

Taxon 1 000   Taxon 1 000
Taxon 2 100   Taxon 2 100
Taxon 3 110   Taxon 3 110
Taxon 4 111   Taxon 4 111
Taxon 5 111   Taxon 5 ???
Taxon 6 ???   Taxon 6 111

```
        ┌──── Taxon 1
       ┌┴──── Taxon 2
      ┌┴───── Taxon 3      MF = 1.0
     ┌┴────── Taxon 4
    ┌┴─────── Taxon 5
    └──────── Taxon 6
```

Taxonomic Congruence = 0   Character Congruence = 1

Figure 5.3 Topological vs character incongruence. One missing taxon in sets A and B causes zero taxonomic congruence (as measured by strict consensus) but complete congruence at the character level.

Discussion

Although we can never measure whether real data sets are accurate or not, we can assess their consistency. By definition, accurate methods will be convergent. Most current discussions of congruence are based on one of two notions of agreement: taxonomic and character. Taxonomic congruence (*sensu* Mickevich and Farris, 1981) measures the similarity in the topologies of systematic hypotheses. This has the strength of addressing the similarity of historical conclusions among data sets, but is very sensitive to small shifts in a single taxon. As has been shown many times (e.g. Wheeler, 1995), a single unstable taxon can reduce taxonomic congruence to such low levels that meaningful comparisons are impossible. Additionally, data sets must contain precisely equal taxon samples to be comparable (pruned comparisons aside). If some taxa are not available for analysis by certain data sets (e.g. extinct taxa and molecular information), taxonomic incongruence is not easily measured. Character-based incongruence, on the other hand, avoids these problems (Fig. 5.3). In character congruence, the extra homoplasy required by combining data is used to assay the agreement among characters. Taxa without data from a particular source are no problem, they are merely missing entries. This metric (Mickevich and Farris, 1981) (MF) has been used frequently to measure congruence (Kluge, 1989; Wheeler *et al.*, 1993; Wheeler, 1995; Whiting *et al.*, 1997). However, character-based incongruence is not without criticism. When comparing phylogenetic methods (or even assumptions), the MF value measures extra homoplasy as a fraction of total character change. This is accomplished by summing up the lengths of the constituent data sets when analysed independently and subtracting that number from the length of the cladogram calculated from the combined data.

This number of extra steps is then normalized via division by this same length of the cladogram derived from the combined data. Hence, MF is the fraction of the total length of this combined data cladogram that is due to combining the data. The greater the difference between the individual analyses and the combined, the greater is the incongruence. Suppose, however, that a poor method is used. In this case, so much homoplasy is already present in the separate analyses that combining them changes little. The number of extra steps and the MF value are low. An example of this would be to use cursory searches for individual data (say without branch swapping) and diligent, time-consuming (or even exact) searches for the combined data. The MF values would be artificially decreased since the individual tree lengths would be artificially long. At present, it is unclear how to deal with this problem except to suspect MF values whose base cladogram lengths are vastly different or perhaps to adjust the MF numbers to reflect the maximum possible incongruence. The rationale behind this adjustment would be that bad methods would express a greater fraction of the possible incongruence in the values used to calculate MF.

Conclusions

Desirable phylogenetic methods are consistent. Although we may not ever know whether an answer is correct ('true') or not, we can assay methods by the constancy of pattern they derive for the same taxa based on different data. Although the measurement of congruence is far from zipless, it does afford us a metric to gauge the behaviour of different reconstruction procedures.

The cases examined here compared nuclear, mitochondrial sequence characters and morphology – a total of six different sources. These data were presented in three different systematic areas and yet demonstrate a consistent pattern of greater congruence for direct optimization than for multiple sequence alignment. If consistency is our guide, direct optimization is superior.

There are several caveats, however, the most important being the use and measure of congruence. The argument has been made (Goloboff, personal communication) that congruence (or at least MF character congruence) cannot be used because it will favour methods that amplify homoplasy. As discussed above, if homoplasy is put in the analysis early, there is little left over to create incongruence when data are combined. This scenario would require that the raw cladogram lengths from which MF is calculated be very different. The 'bad' method would have to have much longer constituent cladograms and similar combined length results (hence lower MF). The direct optimization has *shorter* constituent cladograms. Additionally, the differences in length are in the 10–20% range, not hugely variant (see Table 5.1).

Secondly, no general conclusion about a method as complex as multiple sequence alignment could ever be resolved based on a single implementation. These results will have to be verified with other alignment methods and other data. Even with these reservations, however, the pattern remains. Direct optimization results are more congruent than those derived from sequence alignment.

Acknowledgements

I would like to thank and acknowledge the contributions of Robert Scotland, Mario de Pinna, Andrew Brower, Rob DeSalle, Gonzalo Giribet, Daniel Janies, Amy Litt, and Pablo Goloboff, without whose assistance this work would not have been possible.

References

Farris, J.S. (1970) A method for computing Wagner trees, *Systematic Zoology*, **34**, 21–34.

Fitch, W.M. (1971) Toward defining the course of evolution: minimum changes for a specific tree topology, *Systematic Zoology* **20**, 406–416.

Gladstein, D.S. and Wheeler, W.C. (1997) *POY: the Optimization of Alignment Characters* program and documentation, New York. Available from ftp.amnh.org/pub/molecular.

Goloboff, P. (1996) *PHAST version 1.0* program and documentation. Available from the author.

Kluge, A. (1989) A concern for evidence and a phylogenetic hypothesis for relationships among *Epicrates* (Boidae, Serpentes), *Systematic Zoology* **38**, 1–25.

Mickevich, M.F., and Farris, J.S. (1981) The implications of congruence in *Menidia*, *Systematic Zoology*, **30**, 351–370.

Needleman, S.B. and Wunsch, C.D. (1970) A general method applicable to the search for similarities in the amino acid sequence of two proteins, *Journal of Molecular Biology* **48**, 443–453.

Sankoff, D.D. and Rousseau, P. (1975) Locating the vertices of a Steiner tree in arbitrary space, *Mathematical Progress* **9**, 240–246.

Vrana, P.B., Milinkovich, M.C., Powell, J.R. and Wheeler, W.C. (1994) Higher relationships of the arctoid Carnivora based on sequence data and 'total evidence', *Molecular Phylogenetic Evolution*, **3**, 47–58.

Wheeler, W.C. (1995) Sequence alignment, parameter sensitivity, and the phylogenetic analysis of molecular data, *Systematic Biology*, **44**, 321–332.

Wheeler, W.C. (1996) Optimization alignment: the end of multiple sequence alignment in phylogenetics? *Cladistics*, **12**, 1–9.

Wheeler, W.C. (1998) Alignment characters, dynamic programming, and heuristic solutions, in Schierwater, B., Streit, B., Wagner, G.P. and DeSalle, R. (eds) *Molecular Approaches to Ecology and Evolution*, 2nd edn. Basel: Birkhäuser Verlag, pp. 243–251.

Wheeler, W.C., Bang, R. and Schuh R.T. (1993) Cladistic relationships among higher groups of Heteroptera: congruence between morphological and molecular data sets, *Ent. Scand.*, **24**, 121–138.

Wheeler, W.C. and Gladstein, D.G. (1992) *Malign: a multiple sequence alignment program*, program and documentation, New York. Avaliable from ftp://ftp.amnh.org/pub/molecular.

Wheeler, W.C. and Gladstein, D.G. (1994) Malign: a multiple nucleic acid sequence alignment program, *Journal of Heredity*, **85**, 417.

Wheeler, W.C. and Hayashi, C.Y. (1998). The phylogeny of the chelicerate orders, *Cladistics*, **24**, 173–192.

Whiting, M. and Wheeler, W.C. (1994) Phylogenetic position of the Strepsiptera: evidence for a homeotic reciprocal thoracic transformation *Nature*, **368**, 696.

Whiting, M.F., Carpenter, J.C., Wheeler, Q.D., and Wheeler, W.C. (1997) The Strepsiptera problem: phylogeny of the holometabolous insect orders inferred from 18S and 28S ribosomal DNA sequences and morphology, *Systematic Biology*, **46**, 1–68.

Chapter 6

'Cryptic' characters in monocotyledons: homology and coding

Revisiting old characters in the light of new data and new phylogenies

Paula J. Rudall

Introduction

As many frustrated taxonomists working on Lilianae (Liliaceae *sensu lato*) and other monocotyledons have acknowledged (e.g. Cronquist, 1981), gross floral and inflorescence characters are often of limited systematic value above the genus level in these groups. Although a few families are well defined by floral characteristics (e.g. three stamens in Iridaceae; apical anther dehiscence in Tecophilaeaceae), many Lilianae have relatively unmodified flowers with six tepals and six stamens, and either a superior or inferior ovary. On the other hand, some micromorphological characters, such as microsporogenesis type and tapetum type, have proved to be significant at the family and order levels (e.g. Rudall *et al.*, 1997; Furness and Rudall, 1998). The main problem with these characters is that they are cryptic: they are relatively inaccessible since they require specialist techniques and an understanding of the structures involved. There is always a need for good comparative and developmental data before scoring data matrices, but for micromorphological characters data are sometimes scored from the literature by non-specialists. It is also essential to avoid using 'types' of structures based on taxon names, which sometimes mask more than one character.

This chapter presents selected examples of cryptic characters that are significant at different taxonomic levels within monocotyledons. Character homologies are reassessed with respect to new comparative data and hypotheses of relationships provided by other (mainly molecular) data sets. The topology achieved by a combined molecular/morphological analysis of monocotyledons in general (Fig. 6.1) (Chase *et al.*, 1995b) and refined in subsequent papers, especially on groups within Lilianae (e.g. Rudall *et al.*, 1997), is the framework for re-evaluation of primary homology assessments. The combined tree (Fig. 6.1) was largely congruent with the *rbc*L tree (Chase *et al.*, 1995a), with two major differences: (1) in the combined tree Lilianae formed a monophyletic clade, as opposed to a paraphyletic grade in the *rbc*L tree; and (2) some differences in detail of resolution within the commelinoid clade (see also Rudall *et al.*, 1999). In the morphological tree (Stevenson and Loconte, 1995), some of the major groupings (especially Lilianae) were fragmented and the first-branching monocot taxa were some genera of Dioscoreales, as opposed to *Acorus* in the *rbc*L and combined trees. At first sight, the results of the molecular and morphological analyses were incongruent, but on closer inspection there was considerable support for the combined topology in the morphological matrix,

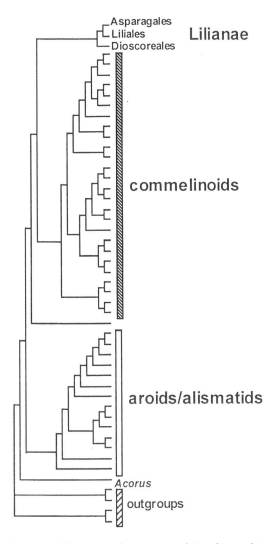

Asparagales
Liliales
Dioscoreales

Lilianae

commelinoids

aroids/alismatids

Acorus

outgroups

Figure 6.1 Topology of monocot relationships achieved by a combined molecular/morphological analysis (Chase *et al.*, 1995b).

and a higher relative level of homoplasy in the *rbc*L dataset than in the combined dataset. Indeed, if the morphological tree was plotted as an unrooted network, with Alismatales at the base, it approximated fairly closely to this topology (Fig. 6.1), with a few taxa displaced (e.g. *Acorus*).

One of the aims of subsequent work has been to review the homoplasy in the morphological data, and to reassess the homologies of some of the critical characters. For example, in the morphological analysis *Acorus* was always linked with *Hydatella* mainly because both were scored as having a perisperm, but on closer examination these were non-homologous (see below).

Characters supporting major clades within monocotyledons: stomatal development

Analyses of *rbc*L and combined data (Chase *et al.*, 1995a, 1995b) supported a major clade Commelinanae (the commelinoid clade), which includes many large groups such as gingers and their allies (Zingiberales), grasses and sedges (Poales) and palms (but not some groups previously associated with palms, such as Pandanaceae and Cyclanthaceae). Commelinanae is well supported by several micro-morphological characters, such as the presence of cell wall ferulates (Harris and Hartly, 1980; Rudall and Caddick, 1994), presence of silica (Rudall and Chase, 1996), *Strelitzia*-type surface waxes (Barthlott and Frolich, 1983; Frolich and Barthlott, 1988), and also stomatal development.

In most morphological matrices, stomata are coded for mature types, especially anomocytic, tetracytic and paracytic, which are the most common mature stomatal types in monocotyledons. Anomocytic stomata lack subsidiary cells, paracytic stomata have lateral pairs of subsidiary cells parallel to the long axis, and tetra-cytic stomata are surrounded by four subsidiary cells, two lateral and two polar (Fig. 6.2) (Metcalfe, 1961). Paracytic and tetracytic forms often occur in combi-nation, and are sometimes scored together. This type of coding provides some taxonomic signal at the genus level, but is relatively 'noisy' at the family and order levels: if the two character states (anomocytic and paracytic/tetracytic) were plotted onto the combined monocot topology (Fig. 6.1), they would both be widely distrib-uted, although the anomocytic type is more common in Lilianae, and paracytic/tetracytic more common in Commelinanae.

However, as Tomlinson (1974) demonstrated, stomatal types in monocotyledons can also be coded as a developmental character, based on whether or not the subsidiary (neighbouring) cells are formed by oblique divisions (Fig. 6.2). Although the anomocytic type, which lacks subsidiary cells, remains the same, the tetra-cytic/paracytic type may be achieved by different ontogenetic pathways, which are therefore non-homologous. For example, in the paracytic and tetracytic stomata of most commelinoids (Poaceae, Cyperaceae, Juncaceae, Centrolepidaceae, Eriocaulaceae, Xyridaceae, Joinvilleaceae, Commelinaceae, Philydraceae, Cannaceae, Marantaceae and Zingiberaceae), the subsidiary cells are derived by non-oblique divisions of the lateral contact cells adjacent to the meristemoid (guard cell mother cell). Conversely, the paracytic and tetracytic types of most non-commelinoid taxa (Agavaceae, some Amaryllidaceae, Asphodelaceae, Butomaceae, Cyclanthaceae, Orchidaceae, Philesiaceae, Pandanaceae) are derived from oblique divisions of the neighbouring cells. This character is therefore highly congruent with the combined (Fig. 6.1) and *rbc*L topologies for monocotyledons, although there are excep-tions: a few commelinoids have oblique cell divisions (e.g. Arecaceae, *Flagellaria*, Heliconiaceae, Pontederiaceae) and some non-commelinoid taxa have non-oblique divisions (e.g. Tecophilaeaceae). Tomlinson (1974) had already pointed out the taxonomic significance of stomatal developmental types at the family level, but recent analyses indicate a further significance at the order level, especially since Pandanaceae and Cyclanthaceae are now not considered closely related to palms (Arecaceae).

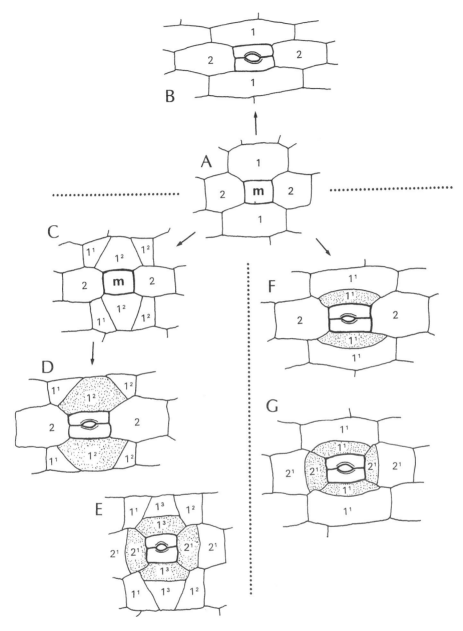

Figure 6.2 Stomata. A, meristemoid and four neighbouring cells. B, anomocytic stoma (without subsidiary cells, i.e. neighbouring cells without derivatives). C–E, development of paracytic
(D) and tetracytic stomata (E) by means of oblique divisions in neighbouring cells. F,
G, development of paracytic (F) and tetracytic (G) stomata by means of non-oblique
divisions in neighbouring cells. m = meristemoid. 1 and 2 = neighbouring cells and their
derivatives (1^1, 1^2, 1^3, 2^1). Redrawn from Tomlinson (1974) with permission.

Characters significant at the order and suborder level: pollen development

Rudall *et al.* (1997), by optimizing data onto a rooted phylogenetic hypothesis derived from analysis of *rbc*L data, demonstrated a correlation between simultaneous microsporogenesis and trichotomosulcate pollen in Asparagales, and discussed the developmental basis for this. In most 'lower' asparagoids microsporogenesis is simultaneous: the second meiosis follows immediately after the first without prior cytokinesis, forming a tetrahedral tetrad within a callose shell (Figs 6.3A and B). In contrast, in 'higher' asparagoids microsporogenesis is invariably successive: cytokinesis occurs prior to the second meiosis, resulting in a tetragonal tetrad (Figs 6.3C and D). Trichotomosulcate pollen, with a forked sulcus, is a variation of the monosulcate type, which is common to most asparagoids, but usually without a forked sulcus. In Asparagales, trichotomosulcate pollen occurs in almost all taxa of the 'phormioid' clade of 'lower' asparagoids (Hemerocallidaceae, or Phormiaceae *sensu lato*; Chase *et al.*, 1996). Not all taxa with simultaneous microsporogenesis have trichotomosulcate pollen; for example, Asphodelaceae is (mainly) simultaneous and sulcate. However, all taxa examined with trichotomosulcate pollen have simultaneous microsporogenesis. There are, therefore, three different groups of asparagoids with respect to these characters: (a) the successive-sulcate 'higher' asparagoids, (b) the simultaneous-sulcate taxa 'lower' asparagoids, and (c) the simultaneous-trichotomosulcate phormioids.

These characters can, therefore, be coded in either of two ways:

1   as two separate characters:

   - microsporogenesis: successive (0) or simultaneous (1)
   - pollen: sulcate (0) or trichotomosulcate (1)

2   as a single character with three unordered character states

   - microsporogenesis: successive, tetragonal (or non-tetragonal) tetrads, sulcate apertures (0), simultaneous, tetrahedral (or non-tetrahedral) tetrads, sulcate apertures (1) or simultaneous, tetrahedral tetrads, trichotomosulcate apertures (2).

Since we have demonstrated a link between these two characters or character states, possession of one (simultaneous microsporogenesis) is a preadaptation to development of the other (trichotomosulcate pollen), and this argues in favour of the second coding option above.

Family characters: ovule structure

Ovule structure, especially the structure of the nucellus, is highly variable in monocotyledons, and yields characters that are relevant at a variety of taxonomic levels. Nucellus characters are sometimes referred to as embryological characters, but in fact relate to sporophytic tissue. Nucellus characters are significant at higher levels in dicotyledons: among eudicots, Asteridae are mainly tenuinucellate and Rosidae are mainly crassinucellate, and this correlates to some extent with number

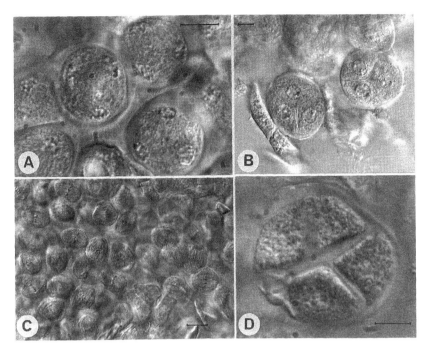

Figure 6.3 Microsporogenesis (differential interference contrast): A and B simultaneous, C and D
successive. A, *Caesia parviflora*, formation of tetrads. B, *Asphodeline lutea*, tetrahedral
tetrads. C, *Hemiphylacus latifolius*, formation of dyads. D, *Thysanotus spinigera*, tetragonal
tetrad. Scale bars = 10 μm.

of integuments (e.g. van Tieghem, 1898; Young and Watson, 1970; Philipson, 1974;
Dahlgren, 1975).

Zuleitungsbahn – postament

With regard to the proximal (chalazal) end of the nucellus, probably the commonest
condition for monocotyledons is one where the nucellus has an extensive subdermal
region (Fig. 6.4A), often with a hypostase. Several proximal nucellus characters are
significant at the family and genus levels in monocotyledons. For example, a highly
differentiated nucellus (Fig. 6.4B) represents the most consistent apomorphy for
Lomandraceae (Lilianae – Asparagales); ovule data are highly congruent with
the *rbc*L tree for this family (Chase *et al.*, 1996). In Lomandraceae the proximal
nucellar region is relatively long and several-layered. Dermal cells are enlarged
and there is little subdermal tissue, but a long central conducting passage
(*Zuleitungsbahn*: Westermaier, 1897) is present, consisting of axially oriented, sometimes darker-staining cells connecting the chalazal vascular supply with the large
antipodals (Fig. 6.4B) (Rudall, 1994). The dermal cells degenerate after fertilization,
leaving the resistant central conducting passage (*Zuleitungsbahn*), which therefore
becomes a postament. Although the postament is sometimes regarded as a (separate)
characteristic feature, at least in this case the postament and central conducting

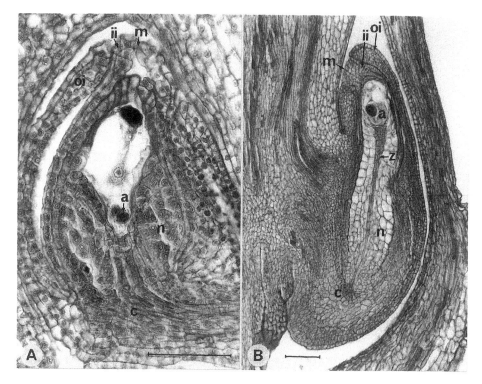

Figure 6.4 Longitudinal sections of ovules. A, *Gilliesia graminea*, proximal region of nucellus (n) short and broad. B, *Lomandra integra*, proximal region of nucellus (n) long and narrow, with central conducting passage (*zuleitungsbahn*) (z). a = antipodals, c = chalaza, ii = inner integument, n = nucellus, m = micropyle, oi = outer integument, z = central conducting passage (*zuleitungsbahn*). Scale bars = 100 μm.

passage (*Zuleitungsbahn*) are homologous structures at different stages of development (Rudall, 1997), and should therefore be coded as the same character.

Perisperm and chalazosperm

A perisperm, which has been recorded in many monocotyledons, is a storage tissue derived from the nucellus. In cases where a perisperm is recorded, an endosperm is often (although not always) absent. However, many instances have been demonstrated where an endosperm is present, but the nucellus still has a role as a regulating or storage tissue for the megagametophyte and/or developing embryo (e.g. in *Asparagus*: Robbins and Borthwick, 1925). Perisperm may develop before anthesis and fertilization.

Both *Hydatella* (Hydatellaceae) and *Acorus* (Acoraceae) are recorded as examples of monocotyledons with a perisperm (e.g. Dahlgren *et al.*, 1985). Indeed, this was one of the main synapomorphies linking *Acorus* with *Hydatella* in Stevenson and Loconte's (1995) morphological analysis of monocot taxa. However, following the results of the combined analysis (Chase *et al.*, 1995b), which conflicted

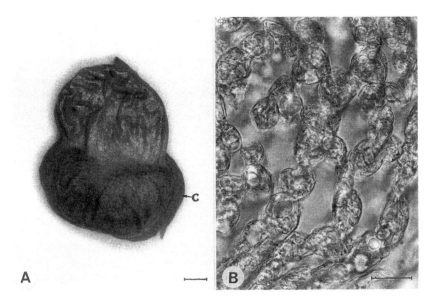

Figure 6.5 *Cyanastrum cordifolium*. A, entire seed (c = chalazosperm; scale bar = 1 mm). B, longitudinal
section, loosely-packed, starch-rich chalazosperm tissue (scale bar = 50 μm).

with this aspect of the morphological data, on critical re-examination it became
clear that the two perisperms are not homologous. In *Hydatella* and most other
taxa (Rudall, 1997) the perisperm is subdermal and multi-layered, whereas in
Acorus (Rudall and Furness, 1997) the perisperm is entirely derived from the single
dermal layer of the nucellus, in which the cells are enlarged and filled with clear,
proteinaceous contents. This character must therefore be recoded as two separate
characters; dermal perisperm is (yet another) autapomorphy for *Acorus*, whereas
subdermal perisperm occurs in several monocot (and dicot) taxa, and is probably
highly homoplastic.

Engler (1901) recorded a perisperm in *Cyanastrum*, but Fries (1919) and Nietsch
(1941) correctly pointed out that this starch-filled loosely packed tissue, (Fig. 6.5)
is not derived from the nucellus, which degenerates after fertilization, but from the
chalaza, outside the raphal bundle. Since endosperm is absent from the mature
seed, Fries (1919) considered that it has a nutritive role for the developing embryo,
and therefore coined the term 'chalazosperm', but conceded that its function is by
no means certain, as it has strong similarities with seed-dispersal structures such
as arils or elaiosomes, which are often found in Asparagales. Chalazosperm is a
tissue derived from the chalaza. A chalazosperm is recorded only in *Cyanastrum*,
and although this is an autapomorphy, *Cyanastrum* was traditionally accorded separate family status (Cyanastraceae) largely on this basis. Analysis of molecular data
from *rbc*L and other genes (e.g. Chase *et al.*, 1995a) supports earlier contentions
(e.g. Hutchinson, 1934) that *Cyanastrum* is embedded in Tecophilaeaceae, with
which it shares many characters (see Brummitt *et al.*, 1998).

Chalazosperm and perisperm are therefore derived from entirely different tissues,
but many authors have continued to homologize them because of the assumed

similarity in function (which in itself is by no means certain). For example, *Eriospermum*, which has a subdermal perisperm (Lu, 1985), has previously some-times been associated with *Cyanastrum* and Tecophilaeaceae, partly on the basis of this character (seed storage tissues), although neither *Cyanastrum* nor Tecophilaeaceae have a perisperm. Analysis of molecular data from *rbc*L, together with other characters such as microsporogenesis (Rudall *et al.*, 1997; see above) have demonstrated that *Eriospermum* belongs in the 'higher' asparagoids, and is not closely related to Tecophilaeaceae.

Conclusions

These examples demonstrate how evaluation of characters in the light of an inde-pendently-derived phylogeny (in this case from *rbc*L) can not only increase our understanding of relationships within groups at different taxonomic levels, but, by reciprocal illumination, can also provide new insights into character homologies. In some instances, separate characters have been mistakenly lumped together; for example, the mature stomatal types 'lump' separate characters, as demonstrated by developmental observations. Perisperm and chalazosperm are often combined as a single character (non-endospermous seed storage tissues), but are morphologically distinct and non-homologous. Perisperm itself represents more than one character: dermal perisperm (*Acorus*) or subdermal perisperm (e.g. *Hydatella*). Some charac-ters are preadaptations for others; for example, trichotomosulcate pollen apparently occurs only in taxa with simultaneous microsporogenesis.

Acknowledgements

I am grateful to Mark Chase (Royal Botanic Gardens, Kew) for help with many aspects of this work, and to Professor P.B. Tomlinson (Harvard University) for permission to use his figure on stomatal development.

References

Barthlott, W. and Frolich, D. (1983) Mikromorphologie und Orientierungsmuster epicutic-ularer Wachse-Kristalloide: Ein neues systematisches Merkmal bei Monokotylen, *Plant Systematics and Evolution*, **142**, 171–185.

Brummitt, R.K., Banks, H., Johnson, M.A.T., Doherty, K., Jones, K., Chase, M.W. and Rudall, P.J. (1998) Taxonomy of *Cyanastroideae* (*Tecophilaeaceae*): a multidisciplinary approach, *Kew Bulletin*, **53**, 769–803.

Chase, M.W., Duvall, M.R., Hills, H.G., Conran, J.G., Cox, A.V., Eguiarte, L.E., Hartwell, J., Fay, M.F., Caddick, L.R., Cameron, K.M. and Hoot, S. (1995a) Molecular systematics of Lilianae, in Rudall, P.J., Cribb, P.J., Cutler, D.F. and Humphries, C.J. (eds) *Monocotyledons: Systematics and Evolution*, Kew: Royal Botanic Gardens.

Chase, M.W., Rudall, P.J. and Conran, J.G. (1996) New circumscriptions and a new family of asparagoid lilies: genera formerly included in Anthericaceae, *Kew Bulletin*, **51**, 667–680.

Chase, M., Stevenson, D.W., Wilkin, P. and Rudall, P.J. (1995b) Monocot systematics: a combined analysis, in Rudall, P.J., Cribb, P.J., Cutler, D.F. and Humphries, C.J. (eds) *Monocotyledons: Systematics and Evolution*. Kew: Royal Botanic Gardens, pp. 685–730.

Cronquist, A. (1981) *An Integrated System of Classification of Flowering Plants*, New York: Columbia University Press.

Dahlgren, R.M.T. (1975) The distribution of characters within an angiosperm system. I. Some embryological characters, *Botanical Notiser* **128**, 181–197.

Dahlgren, R.M.T., Clifford, H.T. and Yeo, P.F. (1985) *The Families of the Monocotyledons*, Berlin: Springer-Verlag.

Engler, A. (1901) Beitrage zur Flora von Afrika XX: Berichte über die botanischen Ergebnisse der Nyassa-Sec- und Kinga-Gebirgs-Exped., Cyanastraceae, *Botanische Jahrbuecher*, **28**, 357.

Fries, T.C.E. (1919) Der Samenbau bei *Cyanastrum*, *Svensk Botanisk Tidskrift*, **13**, 295–304.

Frolich, D. and Barthlott, W. (1988) Mikromorphologie der epicuticularen Wachse und das System der Monokotylen, *Tropische und Subtropische Pflanzenwelt*, **63**, 1–135.

Furness, C.A. and Rudall, P.J. (1998) The tapetum in monocotyledons: structure and systematics, *Botanical Review*, **64**, 201–239.

Harris, P.J. and Hartley, R.D. (1980) Phenolic constituents of the cell walls of monocotyledons, *Biochemical Systematics and Ecology*, **8**, 53–160.

Hutchinson, J. (1934) *Tecophilaeaceae. The Families of Flowering Plants. 2. Monocotyledons*, London: Macmillan, pp. 102–104.

Lu, A.M. (1985) Embryology and probable relationships of *Eriospermum* (Eriospermaceae), *Nordic Journal of Botany*, **5**, 229–240.

Metcalfe, C.R. (1961) The anatomical approach to systematics. General introduction with special reference to recent work on monocotyledons, in *Recent Advances in Botany*, Toronto: University of Toronto Press, pp. 146–150.

Nietsch, H. (1941) Zur systematischen Stellung von *Cyanastrum*, *Oesterreichische Botanische Zeitschrift*, **90**, 31–52.

Philipson, W.R. (1974) Ovular morphology and the major classification of the dicotyledons, *Botanical Journal of the Linnean Society*, **68**, 89–108.

Robbins, W.W. and Borthwick, H.A. (1925) Development of the seed of *Asparagus officinalis*, *Botanical Gazette*, **80**, 426–438.

Rudall, P.J. (1994) The ovule and embryo sac in Xanthorrhoeaceae *sensu lato*, *Flora*, **189**, 335–351.

Rudall, P.J. (1997) The nucellus and chalaza in monocotyledons: structure and systematics, *Botanical Review*, **63**, 14–184.

Rudall, P. and Caddick, L.R. (1994) Investigation of the presence of phenolic compounds in monocot cell walls, using UV fluorescence microscopy, *Annals of Botany*, **75**, 483–491.

Rudall, P.J. and Chase, M.W. (1996) Systematics of Xanthorrhoeaceae *sensu lato*: evidence for polyphyly, *Telopea*, **6**, 629–647.

Rudall, P.J. and Furness, C.A. (1997) Systematics of *Acorus*: ovule and anther, *International Journal of Plant Science*, **158**, 640–651.

Rudall, P.J., Furness, C.A., Chase, M.W. and Fay, M.F. (1997) Microsporogenesis and pollen sulcus type in Asparagales (Lilianae), *Canadian Journal of Botany*, **75**, 408–430.

Rudall, P.J., Stevenson, D.W. and Linder, H.P. (1999). Structure and systematics of *Hanguana*, a monocotyledon of uncertain affinity. *Australian Systematic Botany*, **12**, 311–330.

Stevenson, D.W. and Loconte, H (1995). Cladistic analysis of monocot families, in Rudall, P.J., Cribb, P.J., Cutler, D.F. and Humphries, C.J. (eds), *Monocotyledons: Systematics and Evolution*, Kew: Royal Botanic Gardens, pp. 543–578.

Tomlinson, P.B. (1974) Development of the stomatal complex as a taxonomic character in the monocotyledons, *Taxon*, **23**, 109–128.

Van Tieghem, P. (1898). Structure de quelques ovules et parti qu'on en peut tirer ameliorer la classification, *Journal of Botany (Paris)*, **12**, 197–220.

Westermaier, M. (1897) Zur physiologie und morphologie der Angiospermen-samenknospe, *Beitr. Zur Wissenschaftlichen Botanik*, **1**, 255–280.

Young, D.J. and Watson, L. (1970) The classification of dicotyledons: a study of the upper levels of the hierarchy, *Australian Journal of Botany*, **18**, 387–433.

Chapter 7

Process morphology from a cladistic perspective

Peter H. Weston

Introduction

> Multilingualism remains the source of movement and growth in a civilization. The ability to fill the house of reality, intellect and imagination with different furniture is a great pleasure and a great strength. The strengths of comparison and of contradiction. The ability to draw on the originality or strengths of one to enrich another. But for this to happen, writers and intellectuals must play their role, carrying words, images, emotions and ideas back and forth between languages.
>
> (Saul, 1995: 35)

Different scientific disciplines are like different languages in some respects. They have a common intellectual ancestry but once they cease to share working scientists, they start to progress almost independently of one another, developing their own specialized concepts, theories and vocabularies. One result of this kind of intellectual divergence is that words inherited from a common ancestral research programme often come to have different meanings. An example to which I will return is 'homology', a word which has come to be understood quite differently by systematists, molecular biologists, and, as we shall see, some plant morphologists.

This is not necessarily a bad thing. Although we have to take care when trying to understand the content of other scientific disciplines, conceptual divergence allows the parallel testing of the fruitfulness of new ideas. Intellectual isolation allows innovations to flourish in small academic populations despite what may be the dead hand of constraining orthodoxy in more mainstream branches of science. Such divergence, in itself, does not always lead to the emergence of better concepts. However, comparing and contrasting different ways of looking at the same thing is a useful exercise that may lead to innovation, or at least highlight the limitations of alternative approaches. But for cross-fertilization to occur, we must foster conceptual exchange between scientific disciplines from time to time.

Why examine Sattler's opinions in particular? Firstly, they are interesting because Sattler is a plant morphologist, not a systematist and in botany the research streams of morphology and systematics are independent of one another to a surprising degree. His primary goal is the description of developmental processes, not phylogeny reconstruction.

Secondly, although Sattler criticizes cladistics, he does so from a position of radical heterodoxy rather than entrenched orthodoxy. His main target is not cladistics but the essentialism of classical plant morphology and all that it has influenced. Cladistics is criticized because in Sattler's opinion it owes too much to its roots – it preserves some Aristotelian concepts from the classical morphological tradition that he argues should have been scrapped. Instead of going too far (e.g. Darlington, 1970; Cronquist, 1987), cladistics does not go far enough. An examination of Sattler's ideas thus offers more potential for insight than a debate with defenders of the pre-cladistic status quo.

Few attempts have been made to explore the implications of Sattler's ideas for systematics (Sattler, 1986: chapter 4; 1994; Hay and Mabberley 1994), and none of those discussions, whatever their other virtues, demonstrates a sound understanding of cladistic theory or practice.

Before dealing specifically with the implications of Sattler's writings for cladistics, I describe and interpret the general features of his philosophy.

Sattler's morphology

Structure as process

The central tenet of Sattler's approach is that organisms are not structures that have developmental processes, they *are* developmental processes (Sattler, 1990, 1991, 1993). For instance:

> With few exceptions, it is usually implied that a structure has an existence of its own, thus representing a static framework within which processes operate. For example, saying 'a leaf grows' refers to the growth process within the 'leaf' structure. Hence there is the contrast (or difference) between the process, which is dynamic, and the structure, which as a structural category is static... This contrast (or difference) also manifests itself in our common view of development. It is generally assumed that development (i.e., process) produces a structure. Hence, development is seen as the means and structure as the result. However, any stage of development can also be seen as a structure, a structure that is not static but changes into another structure that changes again, and so on. Even the so-called mature structure is not totally static. Hence, structure is always dynamic and therefore it is not different from process; it is process.
>
> (Sattler, 1991: 708)

A biologist's brief observation of an organism and his or her consequent structural description thus represent, respectively, the perception and representation of an instantaneous morphology or 'snapshot' of arrested development. If time-lapse film footage of organismal development is equivalent to a dynamic morphological description, then each still frame from the film footage is equivalent to a purely structural description. For practical reasons our reconstructions of organismal development are always built up from temporal series of such instantaneous morphologies. Often, these snapshots do not even represent stages of the same organ or organism in cases where fixation is necessary before the biological structures of interest can be observed. In these circumstances, different, serially homologous organs or conspecific organisms are sampled at different developmental stages so as to construct an illusion of continuous development. This methodological limitation should not, however, be confounded with the nature of the developmental process itself, which is truly continuous. In order to arrest an organism's development, we usually have to kill it.

Sattler frequently uses the metaphors of flow and fluidity in explaining his conception of organismal development, but this should not be taken to imply a notion of purely linear or hierarchical causality. He emphasises the importance of feedback loops and causal networks in biological systems (Sattler, 1986: 128–133), and cites the work of experimental biologists who have shown the importance of non-linear causality in development (cf. Alberch, 1985; Roth, 1988). Each organism is a whole process, generated by the interactions of functionally integrated subprocesses including its constituent modules (organs, tissues, cells, organelles, etc.). Although some aspects of development such as cellular proliferation and differentiation are predominantly hierarchical, divergent processes, the resulting hierarchical patterns do not necessarily coincide (cf. Roth, 1994). Processes such as meristem induction in response to translocated hormones do not parallel cellular genealogies. Moreover, morphogenesis is constrained by physical processes operating at all organizational levels from the subcellular to the organismal environment (see e.g. Alberch, 1985; Goodwin, 1994a). Organismal integration involves communication within and between organizational levels.

Sattler's dynamic view of the universe is not new (see e.g. Sattler, 1986: 182–186; Popper, 1972: 144–148), but it does prompt a digression to consider whether any subjects of scientific inquiry are truly static. Sattler's answer to this question appears to be 'no': 'Change is universal, i.e., everything changes' (1986: 182). However, I can think of two static subjects of scientific enquiry, although neither could reasonably be called a thing – the regularities described by natural physical laws and history, including phylogeny.

The static nature of history is not discussed by Sattler, perhaps because his tendency to methodological scepticism of historical science in general (Sattler, 1986: 186–188) and of phylogeny reconstruction in particular (Sattler, 1986: 91–92; 1994: 458) renders the question moot for him. But if time is relative and irreversible, then history must be static. All events are unique (cf. Sattler, 1986: 187).

Process and language

If we accept Sattler's dynamic worldview, we can ask whether his argument with biological orthodoxy is purely semantic rather than methodological. That is, if nothing is static then are all 'things' really dynamic entities by definition? Could we keep him happy by simply changing our descriptive language, without changing the way we do biology? If so, his critique would amount to nothing more than a pedantic quibble.

Sattler certainly thinks that linguistic innovation is desirable. He attributes our tendency to delineate structure from process to 'the noun–verb structure of our common language' (Sattler, 1993: 142) and advocates the development of a new descriptive language for plant morphology that would contain no nouns, only verbs. Although such a language seems implausible, he cites the precedent of Amerindian languages that contain few or no nouns, but are composed instead of nested clauses in which verbs replace nouns (Sattler, 1993).

I suspect, however, that a descriptive language that excludes nouns would prove to be severely handicapped because it would eliminate relative landmarks. Just as everyday language relies on a core of accepted 'primitive' elements (see e.g. Popper, 1972: 258–273), any language of morphology is likely to have to rely on a minimal set of accepted, homologous landmarks and surfaces (i.e. nouns). This has proved to be the case in morphometric analysis of structural variables (see e.g. Bookstein *et al.*, 1982; Bookstein, 1994; McLellan and Endler, 1998), and there is no reason to think that dynamic variables will be any less reliant on a basic relational framework. Indeed, the language that Sattler uses to describe developmental parameters (e.g. Sattler and Jeune, 1992; Jeune and Sattler, 1992) relies on a structural context.

Perhaps a better way to express the main tenet of Sattler's programme is to say that organismal structure and development are two sides of the same coin. Sattler does this by arguing that in lieu of a new descriptive language, we ought to use existing language differently:

> Instead of excluding nouns, I interpret the structural entities to which they refer dynamically... Thus a structural entity is seen as a dynamic system... However, such a system is not endowed with an essence.
>
> (Sattler, 1993: 144–145)

At this point, one might conclude that Sattler is merely playing with definitions, not substantive issues. That conclusion would, however, be a mistake. Interpreting structural entities dynamically is better than treating them as instantaneous morphologies because doing so provides an additional dimension of morphological information. In systematics, for instance, developmental dynamics provide us with information about cladogram rooting (e.g. de Pinna, 1994; Weston, 1994; Bryant, 1997), heterochrony (e.g. Kluge, 1988), the direct manifestation of character transformation (e.g. Patterson, 1982), additional characters and additional information in the establishment of primary homologies (e.g. Endress, 1994). Sattler, however, would probably not endorse any of these as good examples of dynamic morphological interpretation, for reasons that I discuss next.

Morphological variation as a patterned continuum

Sattler's likely objection to the way that cladistic analysis handles development would be to say that it struggles with a pseudoproblem – the categorical or 1:1 homologization of structures. He sees any such categorization as the imposition of artificial divisions across a continuum of morphological variation and advocates the concept of partial homology as a better alternative (Sattler, 1966, 1986: 101–124, 1996). For example:

> Typical classical morphology is categorical. The diversity of plant form is reduced to mutually exclusive morphological categories. In the case of flowering plants and some other taxa, the basic structural categories are root, shoot, stem (caulome), leaf (phyllome) and trichome. Thus any structure encountered

must be either a root, a shoot, a caulome, a phyllome or a trichome (or a combination of any of these): *either-or*, as in Aristotelian logic. . .

(Sattler, 1996: 577)

Sattler's alternative is continuum morphology:

In contrast to this categorical view of typical classical plant morphology, continuum morphology acknowledges gradations between typical structures. . . Thus, a continuum of structures became established. From this point of view, homology is a matter of degree. . . Intermediates are partially homologous to typical representatives of structural categories.

(Sattler, 1996: 577)

This concept of homology resembles that of some molecular biologists who speak of two sequences as being '50% homologous' if they share 50% of their aligned nucleotides. Most cladists would refer to this concept as 'structural similarity' rather than as 'partial homology'. Sattler explains the broad acceptance of classical morphology in terms of the frequency of particular, 'typical' patterns within the continuum. The system of arbitrary pigeon-holes works because the great majority of structures can be slotted into it unproblematically:

It is important to note that the continuum is heterogeneous, which means that some areas (namely those of the typical structures) are denser than others (i.e. those of the intermediate structures). . . Furthermore, it should be recognized that the continuum is dynamic. . . This means that each structure can be seen as a process combination and thus the structural continuum is a continuum of process combinations.

(Sattler, 1996: 578)

Morphological variation among vascular plants is thus seen as a 'patterned continuum', which can be modelled as a polygonal cloud of points in a multidimensional 'morphological space' (Fig. 7.1). Each point in this model represents a 'process combination', a particular organ of a particular plant. The distance between the points represents degree of similarity, reflecting the relative number of shared and unshared developmental processes. The organs of most taxa are represented by points that cluster densely at the vertices of the model. For example, the simple leaf of *Syringa vulgaris* would occupy a position near one corner of the morphological space, where it would be part of a dense cloud of points, each representing a different, 'typical', simple, dorsiventral leaf. All of these, for example, would develop from lateral primordia and show limited apical growth. The shoot of *Syringa vulgaris* would be found near another corner of the space, clustering with other 'typical radial shoots'. All of these would, for example, develop from apical meristems and show unlimited apical growth. The unusual indeterminate pinnate leaf of *Chisocheton tenuis*, however, would occupy a position intermediate between these corners, sharing processes with both typical simple leaves (e.g. development from a lateral primordium) and typical radial shoots (e.g unlimited apical growth). This area in the morphological space would be sparsely filled with points, all representing 'atypical' process combinations.

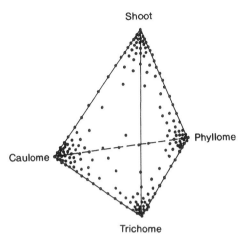

Figure 7.1 Sattler's theoretical model of shoot morphology in which structures from different species are represented as dots in a three-dimensional ordination. The four typical structures that occupy the four corners of the polygon are represented by denser clouds of dots than are intermediates. Redrawn from Sattler (1986, Fig. 5.4a).

Sattler (1986: 105–106) argues that the replacement of the classical categorical model by his continuum model is rationally justifiable in terms of modern philosophies of science. His model explains the success of the classical model in new terms, and in addition explains anomalies that contradict the classical model. It therefore provides excess empirical content (see e.g. Lakatos, 1970).

Sattler's model is not purely theoretical. Sattler and Jeune (1992) and Jeune and Sattler (1992) constructed such a multidimensional model by ordinating organs from a sample of 'typical' and 'atypical' angiosperms. In those studies each structure was characterized by values for eight and thirteen developmental variables respectively and the resulting data matrices subjected to principal component analysis. As predicted, 'typical' roots, stems, leaves, shoots and trichomes clustered at the vertices of a multidimensional polygon. When the values for 'atypical' organs were interpolated into this correlation structure, the additional points occupied intermediate positions mostly within the polygon.

Sattler and Jeune interpreted this empirical result as confirming Sattler's continuum model of plant morphology, and refuting 'the rigidly categorical view' of essentialistic morphologists. This is a fair conclusion, but they did not have to ascend into multidimensional hyperspace in order to argue their case. A morphological continuum is, after all, a logical corollary of evolutionary theory, and the unbroken continuity of phylogenetic lineages going back to the common ancestor of all organisms. Thus the thallus of a liverwort cannot be categorized as either leaf or stem (Jeune and Sattler, 1992) because liverworts probably diverged from other land plants before the evolutionary origin of leaves and stems (Lewis *et al.*, 1997). Similarly, botanists have tended to assume that the organ categories of sepal and petal can be applied to every angiosperm flower that has a perianth. However, some so-called primitive angiosperms such as *Calycanthus* (allspice) and *Idiospermum* have numerous, similar perianth segments that cannot be categorized as either sepals or petals because none of their ancestors are likely ever to have had differentiated sepals and petals (Albert *et al.*, 1998). Morphological intermediacy can, in principle, be explained as a manifestation of shared, primitive characteristics – symplesiomorphy.

Sattler and Jeune refuted only a categorical model of plant morphology in which all categories are mutually exclusive (like a set of pigeon-holes, or a flat-file database structure), not categorical models *per se* (including nested hierarchies). Although Sattler's model completely lacks a taxic dimension, it is not necessarily incompatible with a taxic approach. Sattler and Jeune's ordinations are compatible with a nested hierarchy of categories, i.e. with phylogenetic patterns. The relative density of points in parts of Sattler's patterned continuum can be explained in terms of differential survival and diversification of lineages. Morphologically 'intermediate' process combinations sampled from so-called primitive plants occupy sparsely occupied areas of the ordination space because they represent evolutionary relicts, sporadic survivors of previously more diverse lineages.

Morphological intermediacy and homeosis

Sattler argues, however, that symplesiomorphy is not solely responsible for morphologically intermediate organs in land plants. He proposes an evolutionary hypothesis of 'developmental hybridization', involving recombination of developmental processes, giving rise to novel process combinations. Homeosis, 'the total or partial replacement of one part by another of the same organism' (Sattler, 1994: 438), is the most striking example of developmental hybridization and is the focus of most of Sattler's recent empirical research (Sattler, 1988, 1994; Sattler and Rutishauser, 1990, 1992; Cooney-Sovetts and Sattler, 1986; Lacroix and Sattler, 1994; Lehmann and Sattler, 1992, 1993, 1994, 1996, 1997). Lately, homeosis has received much attention from molecular developmental botanists and zoologists because homeotic mutations are markers for regulatory genes that have a pivotal role in pattern formation (see e.g. Coen, 1991). The best examples of homeosis in plants are 'double-flowering' cultivars in which sepals and/or stamens and/or carpels are replaced by petaloid organs (see e.g. Lehmann and Sattler, 1996, 1997). However, homeosis is not just shown by the flowers of cultivars and mutant model organisms. Many cases of floral and vegetative differences between taxa can also be explained as the result of homeotic change. For example, petals appear to have 'replaced' sepals or stamens and vice versa in a number of plant groups (e.g. Albert *et al.*, 1998; Lehmann and Sattler, 1992, 1993, 1994). Similarly, the 'leaves' of *Asparagus* have been termed 'phylloclades' because although superficially resembling leaves, they develop in axillary positions usually occupied by shoots (e.g. Cooney-Sovetts and Sattler, 1986). In a number of Acacia species such as A. *verticillata*, morphologically anomalous leaf-like organs develop in positions on the stem not usually occupied by any other organs (e.g. Kaplan, 1984; Rutishauser and Sattler, 1986; Sattler *et al.*, 1987).

It is worth looking at an example of postulated homeotic evolutionary change more closely. Lehmann and Sattler (1993) described floral development in the bloodroot, *Sanguinaria canadensis* (Papaveraceae), a species in which the flowers are usually described as having four perianth whorls, with two sepals in the outer whorl, two petals in each of the second and third whorls and four petals in the fourth whorl. The fourth whorl develops from primordia that correspond positionally to stamen primordia in its sister species, the snow-poppy, *Eomecon chionantha* (see Blattner and Kadereit, 1995) and other outgroup taxa. In shape, these primordia also resemble those of stamens rather than those of 'normal' petals. After this stage, however, their developmental trajectory dramatically changes, approaching that of a petal. At anthesis, they differ from both stamens and petals, although they do resemble petals closely enough to have been categorized as such in most morphological descriptions.

Lehmann and Sattler (1993) concluded that the 'ectopic' petals of *Sanguinaria canadensis* are partially homeotic, showing some characteristics of petals and some of stamens. According to Sattler's general theory of developmental hybridization, the ectopic petals show partial homology to both petals and stamens.

Note that this conclusion is a hypothesis of transformational homology (in the sense of Patterson, 1982), which relies on phylogenetic knowledge. But it differs from conventional hypotheses of transformational homology in postulating evolutionary change by recombination of existing developmental processes rather than the generation of a novel developmental process.

Some general comments on Sattler's morphology

Sattler's philosophical justification for preferring process/continuum morphology over structural categorization is based on a distaste for abstractions. For example:

> A comparison of the dynamics of whole organisms would allow us to avoid fragmenting organisms into parts, characters, and character states. Such fragmentation is the first step in the establishment of structural homologies and thus all homologization in the traditional sense is somewhat removed from nature because of the abstraction that leads to fragmentation. . . . Processes are also abstractions. However, they appear less removed from nature than structural fragments. . . Since they interlace, a dynamic view of organisms is less fragmenting than structural decomposition.
>
> (Sattler, 1994: 465)

It is worth making the point here that all scientific knowledge is composed entirely of abstractions – models that simulate reality. There is nothing wrong with abstraction. It is the way that abstractions are used in science that is important. Their usefulness is entirely context-dependent. So it is pointless to state that a model is inadequate in an absolute sense unless it is internally contradictory or syntactic nonsense. A scientific model should be judged on the basis of its functional role in a system of empirical hypothesis-testing. For example, we saw how Sattler and Jeune's multidimensional morphological model could have a legitimate role in the empirical rejection of essentialistic morphological categories. However, since it has no taxic dimension, it is useless, as it stands, as a basis for phylogenetic reconstruction. Similarly, characters and their states may appear to Sattler to be more removed from nature than morphogenetic distances, but they do provide the taxic hypotheses necessary for phylogenetic analysis. In response to Sattler's assertion that the concept of morphological characters is too static, it is reasonable to ask 'too static for what?' More pertinent questions are whether morphological character analysis has helped us to make scientific progress and whether it can be replaced by a more effective alternative.

Sattler's research programme represents a serious attempt to refute the basis of classical morphology and to replace it with a better alternative. In trying to develop a useful, purely descriptive replacement, he has sought to avoid incorporating any evolutionary concepts or theories, at the same time as eschewing idealism and essentialism.

Judged by these standards, the result is a valiant failure, because simultaneously satisfying all of these requirements is, I think, impossible. In avoiding phylogenetic realism (homology as synapomorphy), Sattler strays towards idealism (homology as idealized distances). Rejecting morphological categories (organ types, qualitatively defined characters and character states) also precludes testing the reality and membership of categories of any kind, including taxa. Using structural descriptive terms on the understanding that they really refer to developmental processes does nothing to improve descriptive precision. In a number of respects, Sattler's critique of classical morphology seems to have 'thrown the baby out with the bath water'.

This is not to say that Sattler's efforts have been futile, but rather that his approach has its weaknesses and limitations as well as its strengths. Let us examine some of these more closely.

The idea that biological structures are developmental processes seems to me to provide a reasonable ontology of form and to be one of the strengths of process morphology. But Sattler treats this ontological theory as though it logically implies a methodological prescription – that it would be best to perceive and describe morphology solely in terms of developmental processes; that it would be best to eliminate structural description altogether. However, if structure equals process then does not process also equal structure? Every observation of instantaneous morphology is a pattern that reflects a developmental process. Indeed, it is through the study of instantaneous morphologies that we reconstruct developmental dynamics. If one were to model form purely in terms of growth parameters, assuming that such a goal is possible, one would succeed only in ignoring relevant information.

Morphologists have tended to struggle with the conceptual basis of homology – what morphological variation 'really is'. For example, pre-Darwinian idealistic morphologists thought that structural similarities within and between organisms reflected a platonic ideal, the 'Plan of Creation'. Essentialistic morphologists, on the other hand, have treated variation as if it reflects a set of timeless, rigidly defined pigeon-holes into which organisms and structures fall. Phylogeneticists differ from both idealists and essentialists in adopting a realistic interpretation of morphological variation: it is postulated to reflect the distribution among taxa of inherited evolutionary novelties. Sattler's conceptualization of morphological variation as a dynamic continuum is closest to idealism in that the morphological space that it describes is a pure abstraction, corresponding to no metric or parametric space.

This, however, is not necessarily a flaw from a scientific point of view. What is important is whether this abstraction can be used as an instrument in the empirical testing of hypotheses. We have seen that it can help in refuting the rigidly categorical model of essentialistic morphology. Beyond that, however, its utility seems limited to providing a metaphor for perceived morphological similarity, the starting point for any comparative analysis of form. More interesting biological questions can only be asked by proceeding well beyond this point. Sattler's interest in homeosis and developmental hybridization is a case in point. These ideas are fundamentally evolutionary ones that only make sense in a phylogenetic context.

On that note, let us now examine the logical relationship between cladistic and Sattlerian morphological concepts.

Process morphology and cladistic analysis

Sattler has been generally unsympathetic towards cladistic analysis, a stance that is hardly surprising given Hennig's fundamentally static ontology, exemplified by his concept of the semaphoront:

> ...we should not regard the organism or the individual (not to speak of the species) as the ultimate element of the biological system. Rather it should be the organism or the individual at a particular point of time, or even better, during a certain, theoretically infinitely small, period of its life. We will call this element of all biological systematics, for the sake of brevity, the *character-bearing semaphoront*.
>
> (Hennig, 1966: 6)

Starting from this static extreme, cladistic methodology could only move in a more dynamic direction. Some cladists such as Nelson and Platnick (1981) came to see the basic aim of cladistic analysis differently, as classifying whole life-cycles rather than semaphoronts. This resulted in an almost Sattlerian attitude:

> . . .living things 'are material bodies. Like all material bodies, they are processes'. If organisms are processes, then taxonomists classify not merely individual specimens but individual life cycles. . .
>
> (Platnick and Rosen, 1987: 7)

This ontological change brings with it a new view of cladistic characters, which can be seen, in principle, as dynamic relations that characterize similarities and differences between different life cycles. Rather than referring to developmental stages, characters can be conceptualized as developmental transformations (de Queiroz, 1985).

Which of these approaches has prevailed? In practice, a spectrum of approaches to the observation of morphology has been taken by cladists, ranging from studies in which all character data are drawn from the static interpretation of instantaneous morphologies – preserved specimens – to those in which all species of a study group have been observed as living organisms in captivity and in the wild. I would guess that most botanical studies are based on a mixture of observations of living and preserved organisms. This is probably a reflection of pragmatism rather than principle. Although it is possible to conduct a cladistic analysis based only on information gleaned from static, instantaneous morphologies, observation of living organisms and characterizing them dynamically provides us with a better empirical basis.

Character independence

There is however, a remaining point of tension between Sattler's organism-as-process concept and cladistic methodology. A fundamental assumption of cladistics (and all other methods of phylogenetic analysis) is character-independence (see e.g. Felsenstein, 1982), an assumption that Sattler would see as denying the reality of organismal integration. He does not explicitly discuss the matter, but concern for this problem is implicit in his views.

The assumption of character independence seems, on first acquaintance, to be a patently ludicrous notion. How can different features of the same functionally integrated organism, inherited from ancestors shared with other such organisms, be said to be independent? Clearly, the common language definition of 'independent' – 'separate or disconnected' – is not being used here. Nor is 'independent' being used in its ordinary mathematical sense – 'capable of taking any value without regard to the variation of other quantities', although this is getting closer.

'Independence' is used in a relative rather than absolute sense in comparative biology. The relevant context within which 'independence' is defined is determined by the way in which associations between variables are interpreted (cf. Hillis, 1997). Thus Harvey and Pagel (1991) wanted to be able to partition out the effects of phylogeny in comparative data sets in order to examine residual correlation, which could be interpreted as the result of purely functional relationships. In the case of phylogeny reconstruction, the assumption of independence is almost opposite to the effect that Harvey and Pagel were trying to achieve. It is equivalent to the assumption that cladistic covariation, i.e. hierarchical character congruence, reflects phylogeny rather than other factors.

What non-phylogenetic explanations can be invoked to account for cladistic character covariation? Character correlation – multiple characters showing the same taxic distribution – is explicable as the result of many different phenomena, including synapomorphy (e.g. Hennig, 1966), convergent or parallel evolutionary change in response to recurrent selection pressures of the external environment (e.g. Harvey and Pagel, 1991), and developmental integration (e.g. Sattler, 1986).

The relationship of inclusion – characters that are uncorrelated but hierarchically nested among taxa – has fewer potential explanations. These boil down to phylogeny on the one hand and, on the other, the existence of hierarchically related, complex generic forms. Goodwin (1994b) and Ho and Saunders (1994) are among the most enthusiastic advocates of the latter explanation. Briefly, they argue that the number of potential biological forms is really extremely limited, being tightly constrained by the physics of the developmental process itself. Biological evolution is seen as a process in which lineages 'leap' from one stable, generic form to another. If development is a predominantly hierarchical process, then potential generic forms will be hierarchically related to one another. According to this explanation, the taxic hierarchy, as revealed by morphological variation, does not reflect phylogeny, just shared regularities of ontogeny. A very strong prediction that can be deduced from this explanation is that characters without ontogeny, such as those drawn from DNA sequences, should not be expected to be cladistically congruent with morphology.

That prediction has now been fairly well tested. Several molecular loci that have been used widely in molecular systematic studies encode structural molecules, the functions of which have been intensively studied. The plastid gene rbcL, for example, encodes part of an enzyme that is central to the physiology of photosynthesis. Not surprisingly, nucleotide variation in *rbc*L is highly functionally constrained but it is not known to play any direct role in morphogenesis (Albert *et al.*, 1994). It comes almost as close to providing a truly independent test of morphology-based phylogenies as I can imagine. Although secondary structural relationships within the molecule also violate the assumption of character independence, this problem will tend, if anything, to bias results against congruence with morphology. Nevertheless, analyses of plant relationships based on *rbc*L sequences (e.g. Chase *et al.*, 1993) show substantial congruence with morphology-based cladograms. These and numerous other examples of congruence between molecular and morphological data sets have effectively falsified Ho and Saunders' (1994: 114) prediction that 'groups defined on genetic relatedness are not the same as those defined by morphological/anatomical resemblance'. Cladistic analyses of morphological and molecular data sets do frequently disagree in some respects, and such disagreements may well be due to violations of the assumption of morphological character independence. However, instances in which groupings that are strongly supported by morphology are incongruent with those strongly supported by molecular data are rare (Donoghue and Sanderson, 1992). Such strong disagreements are as likely to be due to the complexities of molecular evolution, such as lineage sorting, as to the inadequacy of morphological methods (see e.g. Doyle, 1992).

One has to conclude that treating morphological characters as independent pieces of evidence is a fruitful procedure for the purpose of phylogenetic analysis. But this heuristic success endorses the idea of character independence as a methodological device, not as an ontological theory. Organisms are not just 'character bearers', even if this is a convenient way to regard them when reconstructing their relationships.

The conceptual basis of transformation

Sattler (1984) first criticized the cladistic concept of homology on a basis that can be briefly paraphrased as follows. The states of a character, although different, are regarded as showing 1:1 correspondence as transformed versions of the plesiomorphic state. However, the postulated transformation is not only unobserved; it is usually, in principle, unobservable; because no direct material transformation of structures ever occurred. The developmental processes and structures to which most characters of multicellular organisms refer are reformed during each generation. So, for example, 'a flower does not give rise to another flower' (Sattler, 1984: 385).

He then went on (Sattler, 1994) to agree with Hay and Mabberley's (1994) argument that the concept of character state transformation involves a category error. Since characters and their states are hypotheses, that is, ideas, they cannot be said to have participated in phylogenetic transformation, a real process. According to this argument, the core of cladistic methodology is based on an ontological mistake.

These criticisms are both aimed at a concept of character transformation based on material change. However, character state transformation has always been treated primarily as a metaphor (see Brady (1994) for a detailed discussion of this idea). Furthermore, character states themselves can be seen as manifestations of the same transformational metaphor. There is no category error because characters,

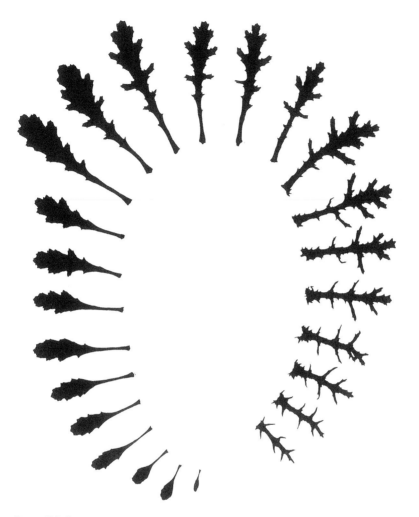

Figure 7.2 Ontogenetic sequence of leaf form in several plants of *Senecio squalidus*, the Oxford
ragwort, from cotyledons (lower left-hand corner) to leaves subtending flowering capitula
(lower right-hand corner).

character states, and character state transformations are all aspects of the same
relation – morphological transformation.

All hypotheses of homology rely on observed morphological *similarities*. The
word 'similarities' is emphasized because 1:1 correspondence in the strictest sense,
absolute identity, is a platonic ideal (cf. Sattler, 1986: 91). Even serially homologous organs such as the leaves of a plant are never individually absolutely identical.
Biological structures are only ever more or less similar to one another. Similarity
is a relation that can be represented by various metaphors such as relative distance,
as in Sattler and Jeune's multidimensional model, or transformation, in which case
dissimilarity is represented by the degree to which the spatial relations within one
structure must be deformed to transform it into another (Brady, 1994).

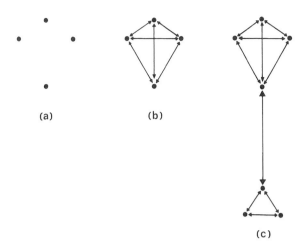

(a)   (b)

(c)

Figure 7.3 (a) Similarities in a comparable feature, represented as relative distances between several individuals (dots); (b) similarities in a comparable feature, represented as figurative transformations (bidirectional arrows) between several individuals; (c) relations within and between two character states represented as a set of figurative transformations (bidirectional arrows) between individuals and between character states.

Consider the analogy of the leaves of a plant, which can be related to one another as members of an ontogenetic transformation series (see e.g. Brady, 1994: Fig. 2.2; cf. Fig. 7.2 here). Progressively interpolating more individual leaves in the series can reduce the extent of transformations between adjacent members to such an extent that an impression of unbroken movement is created, from one end of the series to the other. The leaves can be said to be connected as part of a dynamic relation. But this transformation is metaphorical, not physical, because the leaves have each developed from a different primordium, not directly from one another. A physical transformation in the apical meristem can be inferred from this series but this is a predicted, rather than an observed, phenomenon.

Similarities between different individuals that share a character state are analogous to those shared by serial homologues and can be represented diagrammatically as distances between points in a morphological space (Fig. 7.3a) much like Sattler and Jeunes's morphogenetic model. The transformational relations between individuals in this model can be represented as bidirectional arrows connecting each point (Fig. 7.3b). Like the relations connecting serial homologues in the previous example, these transformations are metaphorical relations or abstractions, not physical processes. Like the serial homologues, we can say that all the individuals sharing a character state are part of a dynamic relation – a transformation.

We can expand this model to include additional points representing individuals possessing another, putatively homologous character state (Fig. 7.3c). The two character states can then be connected by an abstracted transformation just like those connecting the individuals sharing the same character state. All elements of a hypothesis of primary homology can thus be represented as metaphorical transformations, and as parts of the same dynamic relation.

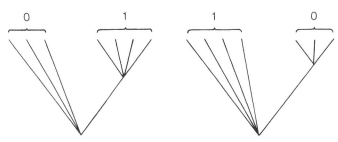

Figure 7.4 The pair of potential taxic relations consistent with the pair of homologous character states shown in Fig. 7.3c.

Are these transformational concepts nothing more than abstractions? Recall that in the analogy of the serial homologues, the figurative transformation linking all the leaves could be seen as predicting an underlying developmental process. The connection between the morphological structures of different individuals is, like that between serial homologues, predicted rather than observed, but in this case it is explained as the product of a shared, inherited, morphogenetic potential. Although molecular developmental genetics is making spectacular progress in elucidating some of the details of those predicted processes (see e.g. Albert *et al.*, 1998), we still know next to nothing about the genetic and epigenetic bases for the inheritance of most morphological similarities.

Cladistic homology and Sattler's partial homology

Sattler's main objection to cladistic analysis is that it relies on 1:1 correspondence of structures in establishing homologies. This approach is at best 'complementary' to his concept of partial homology (e.g. Sattler, 1984: 389, 1994: 463), and at worst forces 'partial correspondences ... into the mold of 1:1 correspondence, which violates and distorts nature' (Sattler, 1994: 462).

The way that I have explained primary homology so far suggests that it is conceptually close to Sattler's concept of partial homology. This is so because both concepts are based on the same 'raw material' – observed similarities between organisms – and the metaphor of distance is a convenient way to represent similarities. There are, however, important differences that distinguish the concepts. Firstly, Sattler's aim is to homologize whole structures, based on the totality of shared and unshared developmental processes, whereas cladistics aims to recognise particular qualitative relations between comparable structures. Secondly, cladistic homology includes a taxic element that is absent from Sattler's concept. Thus the primary homology represented in Fig. 7.3c implies a pair of alternative taxic hypotheses (Fig. 7.4) that can be tested by comparison with other primary homologies under the parsimony criterion.

The distinction that I have drawn between Sattlerian and cladistic concepts of homology resembles Patterson's (1982) distinction between transformational and taxic homology. Sattler's concept differs from Patterson's examples of transformational homology most obviously in lacking any explicit evolutionary interpretation, and in its overt emphasis on developmental dynamics. However, it resembles transformational homology in lacking a taxic dimension. Like transformational homology, partial homology does not generate cladistically testable hypotheses.

Homeosis

Sattler's most interesting objection to all concepts of homology based on 1:1 correspondence, including cladistic concepts, is their supposed inability to deal with the results of homeosis. Consider, for example, the case of accessory ectopic petals in *Sanguinaria canadensis*, to which I referred earlier. Lehmann and Sattler (1993) interpret these ectopic petals as 'hybrid organs' that are not the transformed 'structural descendants' of any one ancestral structure. They combine some processes that take place during the organogeny of stamens and some of petals. While we can interpret them as either petaloid stamens (implicitly giving primacy to the criterion of relative position in formulating homologies) or as displaced petals (implicitly giving primacy to the criterion of composition), if Lehmann and Sattler's interpretation is correct, then our choice of 1:1 homology will be arbitrary. These structures represent a novel process combination that is the result of the recombination of ancestral developmental processes. Any attempt to draw 1:1 homologies between them and any ancestral structures will effectively deny their secondary morphological intermediacy.

Homeosis is not just a problem that can be quarantined to a few specific, obvious examples. If homeosis is both a common evolutionary process and frequently partial in its effects, then we can expect to find a spectrum of secondary morphological intermediacy from almost undetectable instances to examples of major developmental recombination. Does this possibility potentially invalidate the general principle of homologizing character states?

Homeosis has rarely been discussed in the conceptual cladistic literature. Patterson (1982: 48), in his discussion of morphological characters and homology, briefly mentioned homeosis as phenomenon known in insects. He considered it to be 'a relation which fails the conjunction and congruence tests but passes the similarity test [of homology]' and which should therefore be excluded from the list of relations that are of value to the systematist. So one way of dealing with instances of homeosis would be to ignore them as useless from a cladistic standpoint. This, however, would be an *ad hoc* strategy that relied on the systematist's ability to recognise homeosis when confronted by it. If cryptic homeosis is a reality, then rejection is not an option. In any case, I am puzzled by Patterson's opinion that homeosis necessarily fails the congruence test. The presence of ectopic petals in *Sanguinaria* is congruent with other postulated synapomorphies distributed in its cladistic neighbourhood (Kadereit *et al.*, 1994). More generally, there is no reason to think that homeosis is any more prone to homoplasy than other modes of morphological change such as heterochrony. I suspect that Patterson was thinking only of recurrent homeotic mutants in insects when he wrote about homeosis.

Sattler's suggestion that homeosis is a fundamental problem for cladistic analysis is, I think, a misconception based on a failure to distinguish transformational from taxic homology. My understanding of that distinction, which was first drawn by Patterson (1982), is that transformational and taxic homology are the conceptual equivalents in the study of homology to phylogenetic trees and cladograms in the study of relationships. Each cladogram subsumes a set of phylogenetic trees among which cladistic analysis cannot arbitrate. Cladistics is capable of reconstructing only taxic relationships – those of shared ancestry, not ancestral–descendant relationships (see e.g. Nelson and Platnick, 1981). Similarly, each hypothesis of taxic homology subsumes a set of transformational alternatives. Returning to the example of *Sanguinaria*, we saw that ectopic petals could be explained by a variety of transformational hypotheses: they could be petaloid stamens, displaced petals or 'hybrid organs'. To these can be added an explanation derived from Coen's (1991) 'ABC model' of floral whorl identity: an expansion of the field of expression of an A class gene into a whorl in which only B and C class genes were ancestrally expressed. According to this hypothesis, the ectopic petals are novel structures identified by overlapping expression of A, B and C gene functions. The point is that all of these transformational hypotheses are consistent with one taxic hypothesis: that ectopic petals are a synapomorphy characterizing *Sanguinaria*. It does not matter which, if any, we arbitrarily choose as a label for this character.

Before leaving this subject, it is worth noting that homeosis is the result of spatial change (expansion, contraction, loss, migration) in the field of expression of one or more regulatory genes. As such, it can be seen as a special case of a much broader class of evolutionary modifications: spatial change in the field of expression of developmental processes. Within this class, homeosis is only remarkable in having relatively dramatic effects. However, many taxic differences that we find unremarkable, such as differences in distribution of trichomes or pigments over leaf surfaces, also result from spatial changes in developmental fields. Such changes are not usually thought of as compromising the 1:1 homology of the organs concerned. Rather, it is expected that the composition (in terms of developmental processes) of organs will be in a constant, minor state of flux. It is only when such changes result in radically altered process combinations – what zoologists would call the generation of a novel body plan – that we find difficulty in establishing reasonable 1:1 correspondences as working hypotheses.

A good example of such radical change is provided by a group of species of *Utricularia* in which no vegetative structures can be clearly homologized with those of other bladderworts (Sattler and Rutishauser, 1990). In these circumstances, cladists have generally boycotted the offending structures and turned to other, more readily interpreted sources of comparative evidence. Having to ignore morphological variation because it will not fit into the mould of discrete characters and character states is clearly a limitation, but it is one that cladists face more frequently when confronted with continuous variation in the size of structures (see e.g. Stevens, 1991; see Chapter 4). Fortunately, the range of alternative sources of comparative evidence is expanding rapidly, thanks to technological advances in molecular biology, and this limitation is not a major impediment to progress.

Conclusions

I have examined character conceptualization in cladistics by considering the merits of a radical critique – Rolf Sattler's process morphology. The main tenets of Sattler's approach are (1) that organisms are not objects that *have* developmental processes, they *are* developmental processes, and (2) morphological variation among plants is not categorical but continuous. Homeosis is seen as one of the most important processes generating the patterned continuum of plant morphology. Sattler has criticized the cladistic character concept as being 'too static'. He sees cladistic characters as abstractions that fragment the reality of continuous variation into 'either/or' states, which misrepresent dynamic morphogenetic processes. As static abstractions, characters are said to be incapable of undergoing evolutionary transformation.

What can cladists learn from this examination of a radically different approach to morphology? We cannot borrow any concepts or techniques directly, for Sattler (1994) seems to be right in suggesting that process morphology and cladistic analysis are complementary ways of seeing, not competing alternatives.

I have shown that Sattler's criticisms do not impact severely on the ability of cladistics to achieve its primary goal: to reconstruct taxic relationships. However, the counter-arguments that I have used rely heavily on the distinction between taxic and transformational homology. Emphasizing this distinction highlights some of the potential limitations of cladistic analysis.

To me, the most obvious of these is the relatively limited capability of cladistic analysis to reconstruct evolutionary transformations. To the extent that cladistic analysis cannot test transformational homologies, it cannot corroborate them either. This limitation is clearest in cases where we have invented more than one transformational hypothesis to describe a taxic difference, as in the case of ectopic petals in *Sanguinaria*. But the number of plausible transformational hypotheses might be limited to a large extent by our own imaginations. If we can think of only one transformational hypothesis to describe a taxic difference and that character is subsequently resolved as a synapomorphy, it does not necessarily imply that we have accurately modelled an evolutionary transformation.

If we also acknowledge that our ability to reconstruct the distribution of character states among hypothetical ancestors is much weaker than our ability to reconstruct cladogram topologies (a corollary of Farris's (1983) justification of the parsimony criterion), then we have to admit that many of our models of evolutionary change are relatively speculative. I am not suggesting that we should abandon such models – they are one of the most appealing by-products of cladistic analysis – but we should hold them with a lighter hand than well-corroborated knowledge of taxic relationships.

Perhaps we should also be more imaginitive in the way that we map morphological variation onto cladograms. At present, this exercise is usually limited to mapping character states onto nodes, but Sattler is right in saying that decomposing variation into characters is just one way of representing morphological similarities. He also makes a good argument that this procedure may sometimes result in considerable loss or distortion of information. An alternative, more holistic approach that does not seem to have been explored would be the fitting of partial homologies as distances onto cladograms.

Acknowledgements

I am grateful to Rolf Sattler for being generous with his time and reprints over a number of years and for encouraging me to write this review. Thanks also to Jonathon Bennett, Harold Bryant, Andrew Doust, Jim Grimes, Alistair Hay, David

Mabberley, Toby Pennington, Robert Scotland, Peter Stevens and Dennis Stevenson for their helpful, critical comments on earlier drafts of this chapter and/or useful discussions of Sattler's ideas. Of course, none of these people necessarily endorses my opinions.

References

Alberch, P. (1985) Problems with the interpretation of developmental sequences, *Systematic Zoology*, 34, 46–58.

Albert, V.A., Backlund, A. and Bremer, K. (1994) DNA characters and cladistics: the optimization of functional history, in Scotland R.W., Siebert, D.J. and Williams, D.M. (eds) *Models in Phylogeny Reconstruction*, Oxford: Clarendon Press, pp. 249–272.

Albert, V.A., Gustafsson, M.H.G. and Di Laurenzio, L. (1998) Ontogenetic systematics, molecular developmental genetics, and the angiosperm petal. In Soltis, D., Soltis, P. and Doyle, J. (eds) *Molecular Systematics of Plants, II*, Boston, MA: Kluwer Academic Publishers, pp. 349–374.

Blattner, F.R. and Kadereit, J.W. (1995) Three intercontinental disjunctions in Papaveraceae subfamily Chelidonioideae: evidence from chloroplast DNA, *Plant Systematics and Evolution*, 9, 147–157.

Bookstein, F.L. (1994) Can biometrical shape be a homologous character?, in Hall, B.K. (ed.) *Homology, the Hierarchical Basis of Comparative Biology*, New York: Academic Press, pp. 197–227.

Bookstein, F.L., Strauss, R.E., Humphries, J.M., Chernoff, B., Elder, R.L. and Smith, G.R. (1982) A comment upon the uses of fourier methods in systematics, *Systematic Zoology*, 31, 85–92.

Brady, R. (1994) Explanation, description, and the meaning of 'transformation' in taxonomic evidence, in Scotland, R.W., Siebert, D.J. and Williams, D.M. (eds) *Models in Phylogeny Reconstruction*, Oxford: Clarendon Press, pp. 11–29.

Bryant, H.N. (1997) Hypothetical ancestors and rooting in cladistic analysis, *Cladistics*, 13, 337–348.

Chase, M.W., Soltis, D.E., Olmstead, R.G., Morgan, D., Les, D.H., Mishler, B.D. *et al.* (1993) Phylogenetics of seed plants: an analysis of nucleotide sequences from the plastid gene *rbc*L, *Annals of the Missouri Botanical Garden*, 80, 528–580.

Coen, E.S. (1991) The role of homeotic genes in flower development and evolution, *Annual Review of Plant Physiology and Plant Molecular Biology*, 42, 241–279.

Cooney-Sovetts, C. and Sattler, R. (1986) Phylloclade development in the Asparagaceae: an example of homeosis, *Botanical Journal of the Linnean Society*, 94, 327–371.

Cronquist, A. (1987) A botanical critique of cladism, *Botanical Review*, 53, 1–52.

Darlington, P.J. (1970) A practical criticism of Hennig-Brundin 'phylogenetic systematics' and antarctic biogeography, *Systematic Zoology*, 19, 1–18.

De Pinna, M.C.C. (1994) Ontogeny, rooting, and polarity, in Scotland, R.W., Siebert, D.J. and Williams, D.M. (eds) *Models in Phylogeny Reconstruction*, Oxford: Clarendon Press, pp. 157–172.

De Queiroz, K. (1985) The ontogenetic method for determining character polarity and its relevance to phylogenetic systematics, *Systematic Biology*, 34, 280–299.

Donoghue, M.J. and Sanderson, M.J. (1992) The suitability of molecular and morphological evidence in reconstructing plant phylogeny, in Soltis, P.S., Soltis D.E. and Doyle J.J. (eds) *Molecular Systematics of Plants* New York: Chapman and Hall, pp. 340–368.

Doyle, J.J. (1992) Gene trees and species trees: molecular systematics as one-character taxonomy, *Systematic Botany*, 17, 144–163.

Doyle, J.J. (1994) Evolution of a plant homeotic multigene family: toward connecting molecular systematics and molecular developmental genetics, *Systematic Biology*, **43**, 307–328.

Endress, P.K. (1994) *Diversity and Evolutionary Biology of Tropical Flowers*, Cambridge: Cambridge University Press.

Farris, J.S. (1983) The logical basis of phylogenetic analysis, in Platnick, N.I. and Funk, V.A. (eds) *Advances in Cladistics*, vol. 2, New York: Columbia University Press, pp. 1–36.

Felsentein, J. (1982) Numerical methods for inferring evolutionary trees, *Quarterly Review of Biology*, **57**, 379–404.

Goodwin, B. (1994a) Generative explanations of plant form, in Ingram, D. and Hudson, A. (eds) *Shape and Form in Plants and Fungi*, London: Academic Press, pp. 3–16.

Goodwin, B. (1994b) Morphogenetic cascades, generic forms, and taxonomy, in Scotland, R.W., Siebert, D.J. and Williams, D.M. (eds) *Models in Phylogeny Reconstruction*, Oxford: Clarendon Press, pp. 93–111.

Harvey, P.H. and Pagel, M.D. (1991) *The Comparative Method in Evolutionary Biology*, Oxford: Oxford University Press.

Hay, A. and Mabberley, D.J. (1994) On perception of plant morphology: some implications for phylogeny, in Ingram, D. and Hudson, A. (eds) *Shape and Form in Plants and Fungi*, London: Academic Press, pp. 101–117.

Hennig, W. (1966) *Phylogenetic Systematics*, Urbana, IL: University of Illinois Press.

Hillis, D.M. (1997) Biology recapitulates phylogeny, *Science*, **276**, 218–219.

Ho, M.-W. and Saunders, P.T. (1994) Rational taxonomy and the natural system as exemplified by segmentation and phyllotaxis, in Scotland, R.W., Siebert, D.J. and Williams, D.M. (eds) *Models in Phylogeny Reconstruction*, Oxford: Clarendon Press, pp. 113–124.

Jeune, B, and Sattler, R. (1992) Multivariate analysis in process morphology of plants, *Journal of Theoretical Biology*, **156**, 147–167.

Kadereit, J.W., Blattner, F.R., Jork, K.B. and Schwarzbach, A. (1994) Phylogenetic analysis of the Papaveraceae s.l. (incl. Fumariaceae, Hypecoaceae, and *Pteridophyllum*) based on morphological characters, *Botanischer Jahrbücher für Systematik, Pflanzengeschichte und Pflanzengeographie*, **116**, 361–390.

Kaplan, D. (1984) The concept of homology and its central role in the elucidation of plant systematic relationships, in Duncan, T. and Stuessy, T.F. (eds) *Cladistics: Perspectives on the Reconstruction of Evolutionary History*, New York: Columbia University Press, pp. 51–70.

Kluge, A.G. (1988) The characterization of ontogeny, in Humphries, C.J. (ed.) *Ontogeny and Systematics*, New York: Columbia University Press, pp. 57–81.

Lacroix, C.R. and Sattler, R. (1994) Expression of shoot features in early leaf development of *Murraya paniculata* (Rutaceae), *Canadian Journal of Botany*, **72**, 678–687.

Lakatos, I. (1970) Falsification and the methodology of scientific research programs, in Lakatos, I. and Musgrave, A. (eds) *Criticism and the Growth of Knowledge*, London: Cambridge University Press, pp. 91–196.

Lehmann, N. and Sattler, R. (1992) Irregular floral development in *Calla palustris* (Araceae) and the concept of homeosis, *American Journal of Botany*, **79**, 1145–1157.

Lehmann, N. and Sattler, R. (1993) Homeosis in floral development of *Sanguinaria canadensis* and *S. canadensis* 'Multiplex' (Papaveraceae), *American Journal of Botany*, **80**, 1323–1335.

Lehmann, N. and Sattler, R. (1994) Floral development and homeosis in *Actaea rubra* (Ranunculaceae), *International Journal of Plant Sciences*, **155**, 658–671.

Lehmann, N. and Sattler, R. (1996) Staminate floral development in *Begonia cucullata* var. *hookeri* and three double-flowering begonia cultivars, examples of homeosis, *Canadian Journal of Botany*, **74**, 1729–1741.

Lehmann, N. and Sattler, R. (1997) Polyaxial development in homeotic flowers of three begonia cultivars, *Canadian Journal of Botany*, **75**, 145–154.

Lewis, L.A., Mishler, B.D. and Vilgalys, R. (1997) Phylogenetic relationships of the liverworts (Hepaticae), a basal embryophyte lineage, inferred from nucleotide sequence data of the chloroplast gene *rbc*L, *Molecular Phylogenetics and Evolution*, 7, 377–393.

McLellan, T. and Endler, J.A. (1998) The relative success of some methods for measuring and describing the shape of complex objects, *Systematic Biology*, 47, 264–281.

Nelson, G. and Platnick, N. (1981) *Systematics and Biogeography: Cladistics and Vicariance*, New York: Columbia University Press.

Patterson, C. (1982) Morphological characters and homology, in Joysey, K.A. and Friday, A.E. (eds) *Problems of Phylogenetic Reconstruction*, London: Academic Press, pp. 21–74.

Platnick, N.I. and Rosen, D.E. (1987) Popper and evolutionary novelties, *History and Philosophy of the Life Sciences*, 9, 5–16.

Popper, K.R. (1972) *Conjectures and Refutations*. Routledge and Kegan Paul, London.

Roth, V.L. (1988) The biological basis of homology, in Humphries, C.J. (ed.) *Ontogeny and Systematics*, New York: Columbia University Press, pp. 1–26.

Roth, V.L. (1994) Within and between organisms: replicators, lineages, and homologues, in Hall, B.K. (ed.) *Homology, the Hierarchical Basis of Comparative Biology*, New York: Academic Press, pp. 301–337.

Rutishauser, R. and Sattler, R. (1986) Architecture and development of the phyllode–stipule whorls of *Acacia longipedunculata*: controversial interpretations and continuum approach, *Canadian Journal of Botany*, 64, 1987–2019.

Sattler, R. (1966) Towards a more adequate approach to comparative morphology, *Phytomorphology*, 16, 417–429.

Sattler, R. (1984) Homology – a continuing challenge, *Systematic Botany*, 9, 382–394.

Sattler, R. (1986) *Biophilosophy. Analytic and Holistic Perspectives*, Berlin: Spring-Verlag.

Sattler, R. (1988) Homeosis in plants, *American Journal of Botany*, 75, 1606–1617.

Sattler, R. (1990) Towards a more dynamic plant morphology, *Acta Biotheoretica*, 38, 303–315.

Sattler, R. (1991) Process morphology: structural dynamics in development and evolution, *Canadian Journal of Botany*, 70, 708–714.

Sattler, R. (1993) Why do we need a more dynamic study of morphogenesis? Descriptive and comparative aspects, in Barabe, D. and Brunet, R. (eds) *Morphogenese et Dynamique*, Frelighsburg: Editions Orbis, pp. 139–152.

Sattler, R. (1994) Homology, homeosis, and process morphology in plants, in Hall, B.K. (ed.) *Homology, the Hierarchical Basis of Comparative Biology*, New York: Academic Press, pp. 423–475.

Sattler, R. (1996) Classical morphology and continuum morphology: opposition and continuum, *Annals of Botany*, 78, 577–581.

Sattler, R. and Jeune, B. (1992) Multivariate analysis confirms the continuum view of plant form, *Annals of Botany*, 69, 249–262.

Sattler, R., Luckert, D. and Rutishauser, R. (1987) Symmetry in plants: phyllode and stipule development in *Acacia longipedunculata*, *Canadian Journal of Botany*, 66, 1270–1284.

Sattler, R. and Rutishauser, R. (1990) Structural and dynamic descriptions of the development of *Utricularia foliosa* and *U. australis*, *Canadian Journal of Botany*, 68, 1989–2003.

Sattler, R. and Rutishauser, R. (1992) Partial homology of pinnate leaves and shoots. Orientation of leaflet inception, *Botanischer Jahrbücher für Systematik, Pflanzengeschicte und Pflanzengeographie*, 114, 61–79.

Saul, J.R. (1995) *The Doubter's Companion*, London: Penguin Books.

Stevens, P.F. (1991) Character states, morphological variation, and phylogenetic analysis: a review, *Systematic Botany*, 16, 553–583.

Weston, P.H. (1994) Methods for rooting cladistic trees, in Scotland, R.W., Siebert, D.J. and Williams, D.M. (eds) *Models in Phylogeny Reconstruction*, Oxford: Clarendon Press, pp. 125–155.

Chapter 8

Homology, coding and three-taxon statement analysis

Robert Scotland

Introduction

Since the publication of Nelson and Platnick (1991) there has been debate, primarily in the journal *Cladistics* (see Chapter 9), about a new method for analysing systematic data called three-taxon statement (TTS) analysis. The debate surrounding the method has, in part, roots in the distinction made by Hull (1979) between so-called 'pattern cladistics' and the standard approach to cladistic analysis *sensu* Hennig (1966) as interpreted by, for example, Farris *et al.* (1970) and Farris (1983). An issue of discussion concerns the relationship between evolutionary theory, phylogeny and systematics. A second issue concerns the concept of homology and what constitutes systematic data. Until the publication of Nelson and Platnick (1991) issues in relation to pattern cladistics were largely philosophical in nature (Platnick, 1979; Brady, 1985), restricted to differences in terminology (Patterson, 1982), the distinction between trees and cladograms (Platnick, 1977), the role of ontogeny (Nelson, 1978), the justification provided for using cladistic methods (Nelson and Platnick, 1981) and whether evolution is axiomatic for systematics (Nelson, 1989a). Platnick (1985) stated that 'Hennig, Patterson, and myself would all arrive at the same cladogram for any data set we examined'. However, Nelson and Platnick (1991) provide a method for analysing systematic data, based on a taxic view of homology and the information content therein. As such it represents a systematic method that does not rely or seek justification on any aspect of evolutionary theory or presumed knowledge of phylogeny for its implementation. Furthermore, the method often yields different results relative to standard cladistic analysis (Nelson and Platnick, 1991; Siebert and Williams, 1998; Carine and Scotland, 1999). This chapter seeks to explain and explore TTS analysis and the concept of taxic homology *sensu* Patterson (1982, 1988a).

Patterson's (1982, 1988a) classic papers on homology contained certain ideas that are incompatible with the standard method of cladistic analysis. As a prelude to a discussion of these differences, a brief account is now provided of the standard cladistic method for analysing systematic data of the type termed 'paired' (or 'shared' for multistate data) homologues, which are different structures that share topological correspondence.

Table 8.1 contains three homology propositions that Patterson (1982) termed paired (shared) homologues. These three multistate characters, stamen number (1), flower colour (2) and leaf type (3), each have three character states. Table 8.1 also contains two binary characters (4–5) of the type termed 'complement relation'

Table 8.1 Data matrix of five characters shared between eight taxa coded as multistate data including outgroup taxa and binary data excluding outgroup taxa

Taxon	Stamen number	Flower colour	Leaf type	Cystoliths	Retinacula	Coding	Ingroup only
	1	2	3	4	5	12345	123
A	5	Brown	Pinnate	Absent	Absent	00000	
B	5	Brown	Pinnate	Absent	Absent	00000	
C	2	Red	Palmate	Present	Present	11111	000
D	2	Red	Simple	Present	Present	11211	001
E	4	Blue	Palmate	Present	Present	22111	110
F	4	Blue	Simple	Present	Present	22211	111
G	2	Red	Simple	Present	Present	11211	001
H	2	Blue	Simple	Present	Present	12211	011

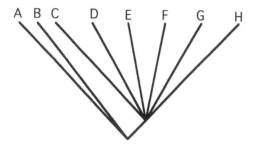

Figure 8.1 Strict consensus tree of seven equally parsimonious trees for the multi-state coding from Table 8.1.

(Patterson, 1982). Assume, for the purposes of discussion, that taxa A and B are outgroup taxa to (CDEFGH). The standard coding protocol for analysing data such as these is provided in Table 8.1. Treating all data as unordered, parsimony analysis for these data yields seven trees with a strict consensus shown in Fig. 8.1. These data are uninformative relative to (CDEFGH). The analysis demonstrates that (CDEFGH) are monophyletic relative to the outgroup taxa A and B. The outgroup taxa contain no homologues (character states) that are found in the ingroup taxa. Omitting the outgroup taxa, the ingroup data can be reduced to binary form as shown in Table 8.1. Analysis of these data yields one unrooted tree (Fig. 8.2a). Fig. 8.2a can be rooted in eight positions, only four of which are characterized by character state changes. The four rooted trees that result from these four positions are shown in Figs 8.2b–e. Inspection of Figs 8.2b–e shows that not a single homologue interpreted as synapomorphy or monophyletic group is constant for all four cladograms.

These two analyses show that multistate data yield an unresolved CDEFGH (Fig. 8.1). Binary data for the ingroup only, yield a single partially resolved unrooted tree (Fig. 8.2a). Comparison of Figs 8.2b–e demonstrates that apomorphic characters are dependent on the position of the root. The unrooted tree (Fig. 8.2a) of unpolarized character state transformations contains no unambiguous groups until the position of the root is imposed. Fig. 8.2a is difficult to reconcile with: 'Every

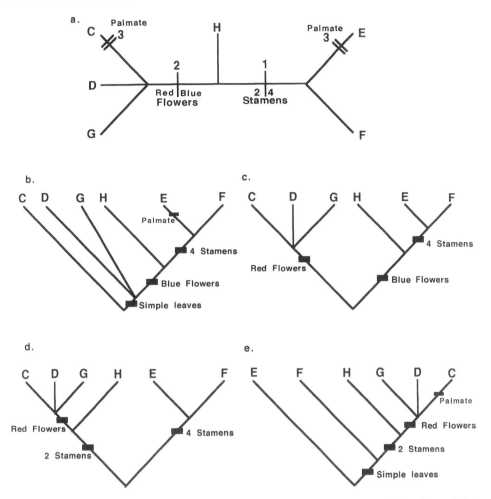

Figure 8.2 (a) Single most parsimonious unrooted tree for the data from Table 8.1 coded in
binary form omitting the outgroup taxa; (b–e) four rooted trees from (a) rooting on
branches characterized by character state changes. Character states shown if diagnostic
of a clade.

hypothesis of homology is a hypothesis of monophyletic grouping' (Patterson, 1982:
33). A question is, do these data contain cladistic structure for resolving taxa
(CDEFGH)? This chapter is an attempt to explore this question.

Patterson's approach

Homology, monophyly and topographic correspondence

'Every hypothesis of homology is a hypothesis of monophyletic grouping' (Patterson,
1982: 33). For Patterson (1982), the shared presence of a particular homologue is
a hypothesis that those taxa that share the homologue are a group. For example,

the shared presence of nucleic acids is a hypothesis that life is a group and is therefore monophyletic. Other taxa such as gnathostomes, vertebrates, mammals, seed plants, angiosperms are similarly characterized by homologues: paired appendages, vertebral column, mammary glands, integumented megasporangia (seeds) and carpels respectively. These homologues are each described by a conditional phrase (Patterson, 1982) which describes the shared attribute. Patterson's (1982) taxic approach to homology contrasts with standard cladistic analysis which views characters as comprising transformation series. From a taxic perspective the homology carpel, at the level of angiosperms, is diagnostic of a group. It is not any particular type of carpel or a carpel characteristic of any particular angiosperm that is the homologue, but the presence of the carpel *per se*.

'The conceptual tool added to, or rather guiding observation in the search for homology is topology' (Rieppel, 1994: 70). 'Homology, as well as topology, are concepts which require the decomposition of organic structures into their constituent elements, and their comparison in terms of connectivity or topological relations within the structural complex' (Rieppel, 1994: 72). Topology is based on connectivity and vice versa (Rieppel, 1988), yet both concepts depend on an invariance of number and relative position of constituent elements in a biological structure (Rieppel, 1988). Riedl (1982: 217) discussed connectivity and topological hierarchy: '. . .it is interesting that the characters of things can be anticipated in a hierarchical arrangement; for example, teeth occur only in jaws and jaws only in skulls. . . If we took the concept of apple, for instance, out of its higher system of tree fruits, fruits, reproductive organs, plants, organisms, it might be an Adam's apple or the apple of your eye'. The frame of reference of topological connectivity is an abstract concept and not unproblematic (see Rieppel, 1988: 44–48, for discussion) but none the less provides the starting point of all worthwhile hypotheses of homology. 'It may, indeed, be true that the origins of hypotheses are immaterial to their scientific status, but it is also true that topological relations of similarity have historically led to greater success in the search for a natural (hierarchical) system of groups within groups than have other premises' (Rieppel, 1994: 70). Therefore homology is a hypothesis of grouping based on topological correspondences. Homology, in this sense, is a hypothesis of hierarchy in that homologies partition organisms into groups. This view is incompatible with the characterization of the standard view of cladistic analysis given above (Figs 8.2a–e) as hierarchy, groups and hypotheses of relationship, in that system, are reliant on the imposition of the root of the cladogram.

Transformational and taxic homology

'Eldredge (1979) distinguishes two approaches to evolutionary theory which he calls "taxic" and "transformational". The first is concerned with the origin of diversity; the second with the process of change. The same distinction can be applied in the study of homologies' (Patterson, 1982: 34). In a similar, vein Rieppel (1994: 90) states that 'homology is not a transformation of an ancestral to a descendent character state. Instead, homology is a unique and deviant condition of form, diagnostic of the particular taxon in question'. Rieppel's (1994: 89) example concerns wings and forelimbs: 'bird wings never pass through the stages of salamander, frog or

Table 8.2 A complement relation homology coded using standard binary notation with a zero (0) = absent and a one (I) = presence

Taxon	Homology I	Coding
A	absent	0
B	present	I
C	present	I
D	present	I

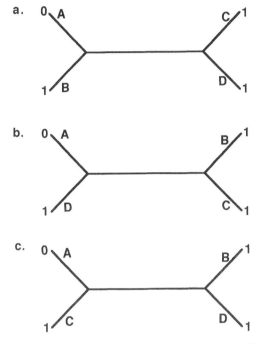

Figure 8.3 Three unrooted trees for the data in Table 8.2 treated as transformation of character states.

lizard forelimbs. Bird wings are always and only bird wings... Deviation from early ontogenetic stages provides the basis for taxic homology'. Thus the taxic view of homology is a hypothesis of a taxon based on the shared homology. The difference between the two approaches, relative to the standard view of cladistic analysis, can be made clear with reference to a four-taxon problem. Table 8.2 is a matrix, in which (BCD) share a homology which is absent in A. Viewed from the perspective of transformational homology, the homology has two character states which are coded zero (0) and one (1). Figs 8.3a–c show the three possible unrooted trees for the four taxa. Coding these data as character states (0: absence) and (1: presence) treats absence as a character state, and the character is interpreted as a one-step character state change in all three trees. All three solutions are therefore equally parsimonious. From a taxic perspective, however, these data imply a single solution: (BCD)A as shown in Fig. 8.4.

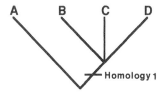

Figure 8.4 The data from Table 8.2, viewed from a taxic perspective, yields an unambiguous solution.

Characters and character states

'I find the distinction (between characters and character states) neither necessary nor useful. The essence of systematics is hierarchy, and in a hierarchical framework homologous "characters" and their "states" represent characters but at more and less inclusive levels... So in what follows, no distinction is necessary or intended between characters or features and their states' Patterson (1988a: 604). Patterson's (1988a) view of characters and character states logically follows from his views relating to taxic homology and treating homology as a hypothesis of monophyly. These views are in contrast to the standard view of cladistic analysis as understood in the writing of Hennig (1966), who considered character states (conditions) of the character as representing a transformation series. Hennig (1966: 89) stated 'but we must always be aware of the fact that "characters" that can be compared are basically only character conditions that the real process of evolution produced by transformation of an original condition'. Hennig (1966) viewed character conditions as ancestral (plesiomorphic) and derived (apomorphic), i.e. encompassing ancestor–descendant relationships. This view is incompatible with Rieppel's (1994: 90) view that: 'homology is not a transformation of an ancestral to a descendent character state'. The taxic and transformational views of homology are inextricably linked to whether a pair of homologues (for example, two and four stamens) are viewed as transformations and therefore coded as two character states of the transformation or whether, in this case, each homologue is viewed as a separate hypothesis of homology. For the data 'two and four stamens', any homology proposition uniting two and four is uninformative at the level of all taxa with two or four. At a more restricted level there are, from a taxic perspective, only two possible hypotheses of homology: the presence of two stamens and the presence of four stamens. In a sense, both the taxic and transformational views of these data achieve the same aim as they both partition the taxa into having two or four stamens (hence the term 'paired homologues'). A difference in the two approaches is that the implication for grouping in the transformational view is reliant on the root of the cladogram. In contrast, the taxic approach views the presence of two stamens as a hypothesis of homology and the presence of four stamens as a hypothesis of homology. Taxic homologies hypothesize taxa and are therefore rooted propositions. Transformational homologies hypothesize unrooted character transformations.

Ancestral characters

Although ancestral (paraphyletic) taxa were eschewed by cladists, ancestral characters (plesiomorphies) have persisted in explanations and the implementation of cladistic methods (Nelson, 1994). Most evolution textbook discussions of cladistic methods (Futuyma, 1986; Ridley, 1996) and reviews of character polarity (Stevens, 1980; Kitching, 1991) view character polarity and the subsequent distinction between ancestral and derived characters as a central feature of a transformational view of cladistic characters. Why then, is it problematic for cladistic characters and their states to be treated: 'as if they, too, were related by descent with modification' (Nelson, 1994: 127)? The first reason relates to the 'hypothetical or conjectural nature of the relationship' (Patterson, 1982: 29) of homology that 'exists in the mind and not in nature' (Nelson, 1970: 378) and although 'not observable, nevertheless [is] rooted in empiricism' (Rieppel, 1994: 89). Rieppel (1994) cited the example of the tetrapod limb which is not an observable structure but an abstraction based on topological relations. Such abstract relations based on topological correspondences between organisms cannot therefore evolve. The tetrapod limb cannot evolve into a wing, as wings are themselves tetrapod limbs. Similarly, the pectoral fins of fishes cannot evolve into tetrapod forelimbs, just as fishes cannot evolve into tetrapods because our current understanding of relationships among gnathostomes shows that some fishes are more closely related to tetrapods than other fishes. For some this argument may seem rather semantic and unimportant and has only served to provide fuel for the creationist literature (Dawkins, 1986: 284). For systematists it remains a central area of discussion since the concept of homology is the pivotal concept of systematics.

Paraphrasing Sattler (1984), Weston (Chapter 7, this volume) states that 'The states of a character, although different, are regarded as showing 1:1 correspondence as transformed versions of the plesiomorphic state. However, the postulated transformation is not only unobserved; it is usually, in principle, unobservable, because no direct transformation of structures ever occurred. The developmental processes and structures to which most characters of multicellular organisms refer are reformed during each generation. So for example, "a flower does not give rise to another flower" (Sattler, 1984: 385). Weston concludes that transformation of character states in a cladistic context is a metaphor.

To question the transformation of character states is simply to entertain the possibility that topographically correspondent structures may not be part of a transformation series (or perhaps never were; see Chapter 7). For example, Fig. 8.5a is an unrooted tree for four taxa supported by 50 paired homologues each of which is distributed (AB)(CD). The 51st pair of homologues (x and x′) are distributed (AC)(BD), shown here as Fig. 8.5b.

The standard approach would treat the non-homology in one of two ways. Either homologue x present in (AC) would give rise to a two-step change in (B) and (D) as shown in Fig. 8.5c or x′ would give rise to a two-step change in (A) and (C) as shown in Fig. 8.5d. However, it is possible to view this conflict between data from a different perspective. Fig. 8.6a is the identical four-taxon problem with 50 paired homologues supporting the tree (AB)(CD). Adding the 51st pair of homologues, x and x′, specifying (AC) and (BD) respectively as shown in Fig. 8.6b, demonstrates that the distribution of x and x′ is incongruent with all other data. An explanation for this is that perhaps x ≠ x, x′ ≠ x′ (Fig. 8.6c). This explanation is precluded by the standard transformational view of cladistic characters because it is treated as unparsimonious. However, this argument only holds within

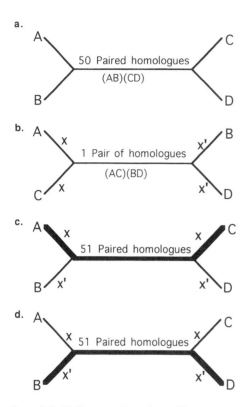

Figure 8.5 (a) Tree resulting from 50 paired homologues with identical distribution between four taxa; (b) 51st paired homologues with incongruent distribution; (c) homologue x interpreted as two-step change to x′; (d) homologue x′ interpreted as two-step change to x.

the assumption of character transformation. If that assumption is relaxed then these data can be interpreted in the context of all other data, which in this case indicates that both character states involve non-homology. A biological explanation of this is that of multiple substitution in molecular data. In that situation topographically correspondent structures are not transformations one of another because multiple substitutions mask the route of the transformation. Therefore, treating data from a non-transformational perspective, rather than being antagonistic to phylogeny reconstruction, attempts to provide a more general system of cladistic analysis that does not rely on the particular model of character evolution provided by Hennig (1966). Standard cladistic analysis tests whether a given transformation occurs once or more than once; the transformation between character states is never questioned (Pleijel, 1995).

Hennig's main contribution to systematics was the discovery of paraphyletic groups (Nelson, 1989b). Hennig (1966) argued that paraphyletic groups were only characterized by plesiomorphies (ancestral characters). Plesiomorphic characters are uninformative in the search for monophyletic taxa. However, plesiomorphic characters in standard cladistic analysis are not distinguished until the tree is rooted after the analysis. The analysis of data therefore includes plesiomorphic characters

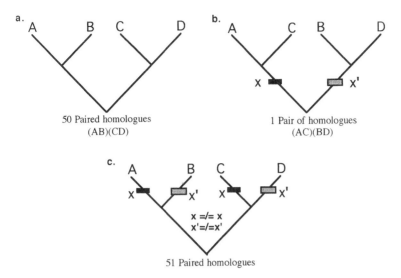

Figure 8.6 (a) Tree resulting from 50 paired homologues with identical distribution between four taxa; (b) 51st paired homologue with incongruent distribution; (c) Most parsimonious tree showing that the incongruent data can be interpreted as non-homology rather than through the lens of plesiomorphy and apomorphy.

and hence cladistic trees can be partly constructed using plesiomorphic characters (see section on character polarity below). This can be understood relative to the following example. Table 8.3 is a matrix for five taxa scored for seven paired homologues (characters 1–7, coded 0 and 1) and four complement relation homologies (characters 8–11, coded 0 = absence, 1 = presence). In this case, we wish to test the putative monophyly of taxa A and B. To test the monophyly of (AB), related taxa C, D and E are included. Standard parsimony analysis of these data yields two unrooted trees (Figs 8.7a and b). For Fig. 8.7a it is possible to root the tree between (AB) and other taxa, but for Fig. 8.7b it is not. The monophyly of AB is therefore equivocal. Consider an analysis of only the paired homologue characters (1–7). These data result in the same two trees (Figs 8.7a and b).

Table 8.3 Data matrix for five taxa with seven paired homologues (characters 1-7) and four complement relation homologues (characters 8-11). For characters 1-7 the zero (0) and one (1) entries represent data whereas for characters 8-11 the zero (0) entry represents absence

Taxon	Paired homologues	Complement relations
	1234567	8 9 10 11
A	0000000	1 1 1 1
B	0001110	1 1 1 0
C	1110001	0 0 0 1
D	1111110	0 1 1 0
E	1110001	0 0 0 1

a.

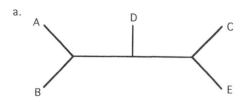

b.

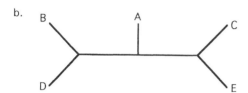

Figure 8.7 Two most parsimonious unrooted trees for the data from Table 8.3.

a.

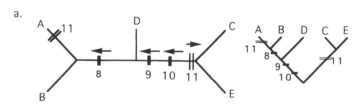

5 STEPS WITH A PARALLEL CHANGE IN HOMOLOGY 11

b.

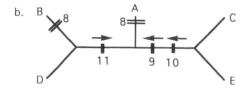

5 STEPS WITH PARALLEL CHANGE IN HOMOLOGY 8 BUT MADE
POSSIBLE AS ABSENCE OF HOMOLOGY 11 IS TREATED AS
CONGRUENT WITH HOMOLOGIES 9 AND 10. ABSENCE OF
HOMOLOGY 11 WOULD CONSTITUTE SECONDARY LOSS TO THE
PLESIOMORPHIC CHARACTER STATE ABSENCE AND WOULD
CONSTITUTE ADDITIONAL HOMOPLASY

Figure 8.8 Two most parsimonious trees for the complement relation homologues from Table 8.3.
(a) (right) Taxic view of these data showing rooted solution compatible with the
unrooted (left).

Consider separately the systematic implications of the four complement relation
homologues (8–11). It is only the presence components of these data that are informative, because absence is not considered a character or a character state (Nelson,
1978; Carine and Scotland, 1999). In ontogeny, absence is more general than
presence (Patterson, 1982). Absence of homologies 8–11 can therefore be considered as plesiomorphic (uninformative). Analysis of these data using the standard

Table 8.4 Data matrix from Table 8.3 showing the hypothetical outgroup coding that assumes no knowledge of polarity in characters 1-7 but treats the absence of data as uninformative for characters 8-11

Taxon	Paired homologues	Complement relations
OG	???????	0000
A	0000000	1111
B	0001110	1110
C	1110001	0001
D	1111110	0110
E	1110001	0001

approach results in the same two trees (Figs 8.8a and b). Figure 8.8a (right) shows the solution of the four complement relation homologues treated from a taxic perspective. However, characters 8–11, coded in binary format and analysed by Hennig86, result in two trees (Figs 8.8a and b). Fig. 8.8b demonstrates that absence and presence are treated as equivalent, and the contradictory nature of Fig. 8.8b cannot be distinguished by an unrooted algorithm. Fig. 8.8b is only possible because the plesiomorphic state (absence) cannot be distinguished in the analysis. A counter-argument could state that these data are best represented as in Table 8.4 with an all-zero outgroup added to polarize the complement data (characters 8–11) and question mark entries (?) for the paired homologue data. Analysis of Table 8.4 does result in Fig. 8.7a and is therefore preferable to the approach that treats absence and presence as equivalent.

Character polarity

Patterson (1982: 52) stated that: 'Initially, all we can say is, if a homologous feature is universal in a group, then the group will be monophyletic. . .'. For a given problem some taxa may have (homologue x) while other taxa have (homologue x′). A homology for all taxa is the presence x + x′ (homologue y). Homologue y unites both, such that x + x′ = y (taxic homology). At the hierarchical level of all taxa,

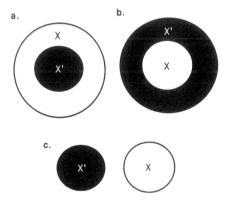

Figure 8.9 A pair of homologues, x and x′, may stand in hierarchical relationship to each other in three ways. (a) x′ is diagnostic of a group and x is not; (b) x is diagnostic of a group and x′ is not; (c) both x and x′ diagnose groups.

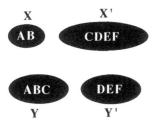

Figure 8.10 Two paired homologues distributed between six taxa.

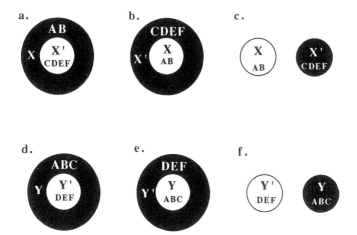

Figure 8.11 The total number of solutions for the data from Figure 8.10. (a) x is plesiomorphic to x′; (b) x′ is plesiomorphic to x; (c) x and x′ both diagnose monophyletic groups; (d) y is plesiomorphic to y′; (e) y′ is plesiomorphic to y; (f) y and y′ both diagnose monophyletic groups.

Figure 8.12 The four informative hypotheses of homology which summarize the data in Figure 8.11.

homologue y is universal and therefore uninformative. Within all taxa only the presence of (x) or/and (x′) can resolve relationships. From a taxic perspective initially all we can say is that homologue x is diagnostic of a group and homologue x′ is diagnostic of a group, as shown in Fig. 8.9c. Patterson's (1982: 35, 51) three possible relations for two paired homologues (x and x′) (Figs 8.9a–c) demonstrate x′ to be informative (Fig. 8.9a), x to be informative (Fig. 8.9b) or x and x′ to be informative (Fig. 8.9c). Figs 8.9a-c are summarized by Fig. 8.9c, because Fig. 8.9c includes Figs 8.9a and b. The informative part of Fig. 8.9a is (x′); the informative part of Fig. 8.9b is (x). Fig. 8.9c includes both and is therefore a general summary of the only two possible informative homologues for these data.

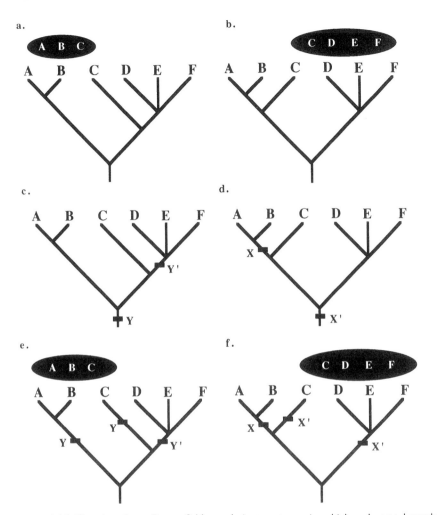

Figure 8.13 The data from Figure 8.11 result in two trees in which only one homologue is not accommodated. (a, b) The two trees with the incongruent homologue shown for each; (c, d) the two trees with the incongruent homologue explained as plesiomorphy; (e,f) the two trees with the informative homologue explained as minimizing the number of uninformative observations.

Fig. 8.10 shows two paired homologues at the level of six taxa. Figs. 8.11a-c summarize the three possible solutions for paired homologues x and x´. Figs 11d-f summarize the three possible solutions for paired homologues y and y'. All six solutions can be reduced to four hypotheses of homology (Fig. 8.12) that summarize the informative relations in all six possible solutions (Figs 8.11a–f). Figure 8.12 results in two trees that are shown in Figs 8.13a and b. Fig. 8.13a contains one homologue (ABC) that does not support the solution. Fig. 8.13b contains one homologue (CDEF) that does not support the solution. How are we to account for these homologues? For homologue (ABC) there are three solutions at the outset (Figs 8.11d–f). Fig. 8.13a excludes the solution in Fig. 8.11e as a possible explanation of these data.

The remaining two solutions (Figs 8.11d and f) are the only two possibilities which can show how to accommodate (ABC) on the tree (Fig. 8.13a). Fig. 8.11d explains (ABC) in the context of plesiomorphy such that although (ABC) is uninformative it fits the base of the tree (Fig. 8.13c) and is interpreted as a precursor for a modified y′ in (DEF). The other possibility (Fig. 8.11f) is that (ABC) is not a modification of (DEF). For this possibility the homologue (ABC) has to be accommodated on Fig. 8.13a. This can be achieved by splitting the homology [homoplasy] such that the (AB) part of the homologue is evidence for a group (AB) but the (C) part of the homologue is incongruent (Fig. 8.13e). This explanation explains all data as congruent (specifying the same tree) except one observation that relates (C) with (AB). This explanation is precluded by standard parsimony analysis on the basis that it is unparsimonious. The solution that treats (ABC) as plesiomorphic for the tree (Fig. 8.13c) contains three observations (ABC) that are uninformative for the tree. Fig. 8.13c contains more uninformative observations than Fig. 8.13e. The problem with evaluating the two possible outcomes is that no matter how badly a homologue fits a tree it can always be accommodated at the base and explained as plesiomorphy with perfect fit (no extra steps). Any alternative explanation is automatically ruled out as it is treated as unparsimonious, even although for the above example, it results in an explanation that has one observation only which is incongruent with the tree. In short, plesiomorphic characters have an exact fit to a tree because they contain no homoplasy. Placing any homologue at the base of a tree achieves a perfect fit no matter how the homologue is distributed among the taxa. For the example there are, in the first instance, four hypotheses of taxic homology (Fig. 8.12). A taxic homology is a hypothesis from a set of observations that result in hypotheses of relationship. Therefore, let's say we wish to minimize the number of observations that are not evidence of relationship. Fig. 8.13e includes one observation (homologue y in taxon C) that is not evidence of relationship. Fig. 8.13c includes three observations (homologue y in taxa A, B and C) that are not evidence of relationship. Fig. 8.13e is a more parsimonious explanation than Fig. 8.13c.

Figs 8.9a–c show the three possible systematic solutions for a pair of homologues. In the context of homoplasy these three solutions are treated unequally. Figs 8.9a and b are preferentially biased because one of the homologues can fit the tree at the base no matter what the distribution of the homologue is between the taxa. Fig. 8.9c, in the context of homoplasy, cannot compete (in terms of counting character state changes) because the explanation will be longer. The three solutions (Figs 8.9a–c) are therefore not competing on an equal level, because in the context of homoplasy, the interpretation of plesiomorphy wins every time. The concept of plesiomorphy can explain any character distribution on a tree without extra steps. Plesiomorphic explanations of character distribution have no competitors. Any homologue which does not fit a tree can be accommodated at the base. In this sense plesiomorphic characters have a substantial influence on the outcome of a parsimony analysis as they serve as a device for excluding any alternative explanations of incongruent data. In this sense, standard cladistic parsimony analysis involves a simple model of character evolution (Nelson, 1994) in which plesiomorphic explanations of data are preferred over any others. Parsimony is applied within the straightjacket of plesiomorphy, so it is not parsimony *per se* that excludes alternative explanations but parsimony working through the lens of a particular model of character evolution.

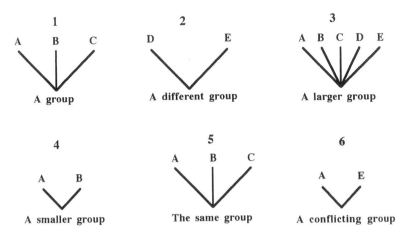

Figure 8.14 Relative to an initial homologue specifying a group (ABC), other homologues can relate to it in any of five ways specifying: a different group (DE), a larger group (ABCDE), a smaller group (AB), the same group (ABC) or a conflicting group (AE).

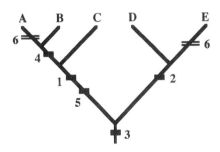

Figure 8.15 Most parsimonious solution for the data from Fig. 8.14.

Character congruence

Patterson's congruence test of taxic homology is often cited as equivalent to finding the most parsimonious solution for a data set within the standard paradigm of cladistic analysis. However, as discussed above, taxic and transformational views of homology seem distinct in that taxic homologies are hypotheses of groups, whereas the standard approach to cladistic analysis treats characters and their states from a transformational perspective. As such, the standard approach to cladistic analysis minimizes character state changes on unrooted trees. However, Patterson's (1988a: 604) taxic view – 'The essence of systematics is hierarchy' – treats hierarchy as inherent in any hypothesis of homology and as such is akin to compatibility methods which are founded on the idea that characters are already hypotheses of relationships (Meacham and Estabrook, 1985). Patterson (1982: 39) showed that there are five ways in which one group specified by a homologue may relate to other homologues (Fig. 8.14). Homologues may specify: a different group from the initial homologue, a larger group that includes the initial homologues, a smaller group within that diagnosed by the original homologue, the same group as the initial homologue or a conflicting group to

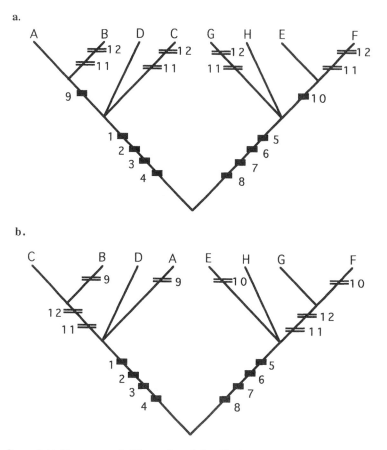

Figure 8.16 Two trees of different length for 12 characters. (a) Length 18, explains characters 1–10 uniquely on the tree; (b) length 16, explains characters 1–8 uniquely on the tree.

the initial homologue. Character conflict can be minimized and character congruence maximized by applying the principle of parsimony which results in Fig. 8.15. However, this approach becomes problematic for other data sets in which minimizing character conflict and maximizing character congruence (non-conflicting characters) are not one and the same (Scotland, 1997; Wilkinson 1997). Fig. 8.16a shows a set of ten characters, 1–10, which are congruent and two characters, 11 and 12, which are incongruent. Fig. 8.16a is length 18. Standard parsimony analysis of these data yields the tree in Fig. 8.16b which is length 16. Fig. 8.16a is the tree discovered by compatibility analysis and maximizes the agreement (congruence) of the largest number of homologies in their original form (cliques), whereas Fig. 8.16b minimizes tree length (Farris, 1983). A difference in these two approaches to congruence amounts to the way in which homoplasy is dealt with in the analysis. Patterson's (1982) similarity and congruence tests of homology (Table 8.5) did not account for characters such as 11 and 12 in Fig. 8.16b. Assuming that these characters pass the similarity test, they are homoplastic but contain a level of synapomorphy as they diagnose groups. Patterson's tests distinguish parallelism and convergence

Table 8.5 Patterson's (1982) similarity and congruence tests of homology

	Convergence	Parallelism	Homology
Similarity	FAIL	PASS	PASS
Congruence	FAIL	FAIL	PASS

Table 8.6 Patterson's (1982) similarity and congruence tests of homology reinterpreted relative to similarity and fit in the context of standard cladistic analysis

	Analogy	Homoplasy	Apomorphy	Plesiomorphy
Similarity	FAIL	PASS	PASS	PASS
Fit	N/A	NOT EXACT $c.i.<1$	EXACT $c.i. = 1$	EXACT $c.i. = 1$
Homology	No	Yes/no	Yes	Yes

c.i., consistency index.

from homology relative to how they perform in the similarity and congruence tests. Here I attempt to recast Patterson's (1982) tests of homology relative to similarity and the concept of fit. Nelson (1994: 110) states that: 'Criteria of homology are useful, even necessary, to organise data but are not decisive in resolving conflict among data during the search for the tree that best fits all data'. Nelson shifts the focus from congruence of characters to best fit between data and all possible trees. Fit in the sense of Nelson (1994) are homologues, in the form of TTSs, that can be accounted for as evidence of relationship. Fit, in the standard cladistic sense, is measured relative to whether a given transformation has an exact fit (consistency index of one) or inexact fit (consistency index less than one). Four categories will be discussed relative to standard cladistic analysis, similarity and fit: analogy, homoplasy, homology and plesiomorphy (Table 8.6).

Analogy

Analogy as used here refers to superficial similarity that is not topographically correspondent; it is often attributed to Owen (1843) but see Panchen (1994). When building a data matrix, analogies fail the similarity test. For example, the wings of birds and insects do not share topographic correspondence and so fail the similarity test. Analogy in this sense is equivalent to convergence *sensu* Patterson (1982). Such characters are pre-screened when constructing a matrix on the basis of comparative anatomy and are therefore not subject to the test of fit. In short, analogies do not take part in the analysis of data.

Homoplasy and homology

At the stage of data matrix construction homoplasy and homology are indistinguishable as they both pass the similarity test of topographic correspondence. This stage is equivalent to de Pinna's (1991) primary homology. The only difference between homology and homoplasy seems to be that homoplasies (characters 9, 10,

Table 8.7 Data matrix for five taxa with two paired homologues (characters 1-2) and one complement relation homology (character 3). For characters 1-2 the zero (0) and one (1) entries represent data, whereas for character 3 the zero (0) entry represents absence

Taxon	Paired homologues	Complement homologues
	12	3
A	00	1
B	00	1
C	10	1
D	11	0
E	11	0

11 and 12 from Fig. 8.16b) do not have an exact fit to a given solution as they have a consistency index of less than one. Homoplasy is a term used to distinguish primary homology statements that have an inexact fit to a given solution. In standard cladistic analysis, homoplasy and homology are relative terms as any given character which contains homoplasy may also contain homology.

Plesiomorphy

Plesiomorphic characters are uninformative in systematics (Hennig, 1966; Farris, 1983). Plesiomorphic characters pass the similarity test but will fail to provide evidence of relationship. In standard cladistic analysis, fit includes plesiomorphic characters as shown in the following example. Fig. 8.17a shows two paired homologues and one complement relation homology for five taxa. Fig. 8.17b shows the most parsimonious solution that results from these data coded as in Table 8.7. The unrooted tree (Fig. 8.17b) has a consistency index of one, indicating an exact fit between the data and the tree. Figs 8.17c and d show two rooted trees from Fig. 8.17b when the root is placed on either of the branches which contain the character state changes. Fig. 8.17d has an exact fit to the solution because the character state 1′ is interpreted as plesiomorphic and is accounted for as fitting the base of the tree even though it does not offer any evidence of relationship between the taxa under study. Fig. 8.17c is made possible because character state 2 is interpreted as plesiomorphic. The absence of homologue 3 is interpreted as evidence for a group (DE). This solution contains two ancestral characters (2 and 3) and one instance of secondary loss (3′). In contrast, Fig. 8.17d contains only one ancestral character and would seem to be a better fit for these data. Nevertheless, in standard cladistic analysis plesiomorphic characters (ancestral characters) are treated as homology with an exact fit to a given solution within the context of character evolution and transformation of character states. In a discussion of paraphyly and plesiomorphy, Nelson (1994: 128) stated that 'Perhaps cladists are fated to repeat the dispute, this time over characters rather than taxa'. In the sense of Patterson (1982) homology equals apomorphy, and therefore plesiomorphy is non-homology (Figs 8.17c and d).

Table 8.6 demonstrates that homoplasy, plesiomorphy and apomorphy are homology.

What distinguishes these three relations is the root of the tree and parsimony optimization procedures. As all three relations contain homology, it is clear that hypotheses of homology are never tested in a standard cladistic analysis.

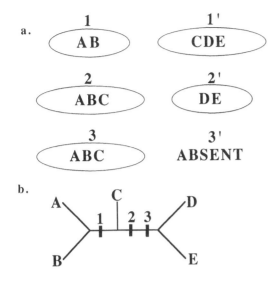

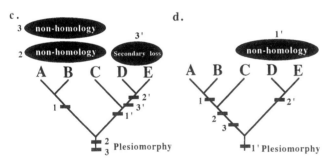

Figure 8.17 (a) Diagrammatic representation of two paired homologues and one complement relation homology shared between five taxa; (b) single unrooted tree for these data; (c) rooted tree from (b) rooted on the branch between (AB) and (CDE); (d) rooted tree from (b) rooted between (ABC) and (DE).

Transformation series and multistate characters

The standard view of cladistic analysis which views homology in terms of transformation of character states (a transformation series) is far removed from the perspective of taxic homology discussed above (Patterson, 1982; Rieppel, 1988; Nelson, 1994; Platnick *et al.*, 1996). Character optimization can involve reconstructing internal nodes on a cladogram in order to discover the tree of minimal length. In the case of multistate characters this can lead to multiple equally parsimonious hypotheses of transformation for a given character (Hawkins *et al.*, 1997). For example, the data in Appendix 8.1 represent DNA sequence data for five taxa, as discussed in the next section. These data yield a total of 106 nucleotide positions which are potentially informative in the context of standard cladistic analysis. From this total, 22 nucleotide positions represent multistate characters. Table 8.8 lists the nucleotide position, pattern of substitution, multistate coding and the three patterns

Table 8.8 The 22 nucleotide positions from Appendix 8.1 that result in multistate characters. Nucleotide position, substitution pattern, coding and pattern of substitution are given

Nucleotide position	Substitutions	Coding	Pattern
121	GCAAA	01222	Three
205	CTACA	01202	Two pairs
211	TAAGA	01121	Three
260	ACTCC	01211	Three
265	CATCT	01202	Two pairs
332	AACTT	00122	Two pairs
352	ACCTC	01121	Three
385	ACCTC	01121	Three
448	CGGAG	01121	Three
452	AGAAT	01002	Three
557	TAAAG	01112	Three
558	ACCTC	01121	Three
567	AGTCC	01233	One pair
611	ACAGA	01020	Three
663	ACTCC	01211	Three
771	GGTCC	00122	Two pairs
780	CAAGA	01121	Three
792	ATCCT	01221	Two pairs
801	ATCTC	01212	Two pairs
822	GCAAG	01220	Two pairs
864	TAAAC	01112	Three
879	AATCC	00122	Two pairs

of character state distribution which are: three taxa sharing same base, two pair of taxa sharing the same base and one pair of taxa sharing the same base. The number and tree topologies, under different character optimizations, for these three patterns are shown in Fig. 8.18. Fig. 8.18 demonstrates that a transformational view of cladistic characters is at odds with the notion of homology as a hypothesis of a group because Fig. 8.18 contains no unambiguous groups until the root of the unrooted tree is imposed. Furthermore, the pattern of substitution at a given position is compatible with several hypotheses of character state transformation which are strongly at odds with Patterson's (1982: 34) assertion that 'The force of a hypothesis of homology is that the inclusive group is monophyletic by virtue of the homology'.

Testing ingroup monophyly and the root of the tree

As outlined above, the standard protocol for testing ingroup monophyly is restricted to rooting the tree between the ingroup and outgroup post analysis. For a five taxon problem there are 105 possible rooted fully bifurcating trees and 15 unrooted trees. If we set out to test the ingroup monophyly of (ABC) relative to D and E, there are three unrooted trees (Figs 8.19a–c) from a total of 15 unrooted trees that are compatible with the monophyly of (ABC). Therefore 1 in 5 (20%) unrooted trees are compatible with (ABC) as monophyletic. However, from the total of 105 rooted trees only nine (8.57%) of these include (ABC) as monophyletic. There are a further twelve trees (Fig. 8.20) that indicate (ABC) to be paraphyletic that are not excluded by the unrooted trees in Figs 8.19a–c.

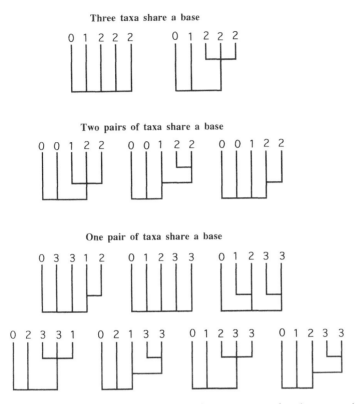

Figure 8.18 Topologies resulting from three patterns of multistate codings from the data shown in Table 8.8.

Therefore the protocol of rooting between the ingroup and outgroup, in this example, includes more trees (12) that are not compatible with ingroup monophyly than trees that support ingroup monophyly (9). The standard practice (Nixon and Carpenter, 1993) of testing ingroup monophyly by constructing unrooted trees and rooting between the ingroup and outgroup is far from being a robust test.

Homology as relationship

Homology is 'the relation that systematists and comparative anatomists use in generating hypotheses of relationship' (Patterson, 1982: 29). Homology, viewed from the perspective of relationship, comprises two distinct elements. First, taxa that share a homologue constitute a hypothesis of a group. Second, taxa that lack a particular homologue are not included in a group hypothesized by that homologue. Homology is a relation between those taxa that share a homologue and those taxa that lack the homologue. For a given systematic problem, a shared homologue for all taxa is uninformative. It is not until a homologue is shared between some taxa but absent in others that a hypothesis of grouping is possible. A hypothesis of homology is a statement of relationship between those taxa that have and those

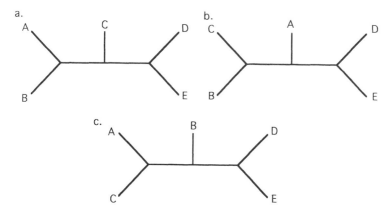

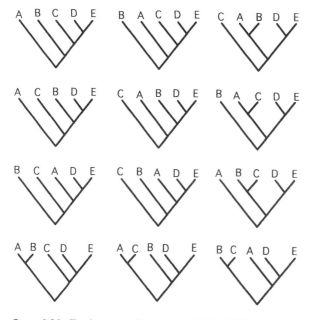

Figure 8.19 Three from a total of 15 unrooted trees for five taxa that are compatible with a monophyletic (ABC).

Figure 8.20 Twelve rooted trees in which (ABC) are paraphyletic that are not excluded by the unrooted trees shown in Figure 8.19.

taxa that lack a particular homologue. Nelson (1994) distinguishes homologues (parts of organisms) and homology (relations between the parts of organisms).

Three-taxon statements

Fig. 8.21a shows a homologue shared between four taxa (ABCD). Given the distribution of the homologue, these data are uninformative. Fig. 8.21b shows the

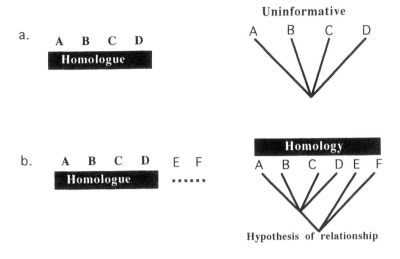

a.
A   B   C   D
Homologue

Uninformative
A   B   C   D

b.
A   B   C   D E F
Homologue ......

Homology
A   B   C   D   E   F

Hypothesis of relationship

c.   Constituent three-taxon statements from (ABCD)EF

(AB)E, (AC)E, (AD)E   (BC)E   (BD)E   (CD)E
(AB)F, (AC)F (AD)F   (BC)F   (BD)F   (CD)F

Figure 8.21 Homologues and homology (a) A homologue shared between four taxa is uninformative; (b) a homologue shared between four taxa but absent in two taxa results in a hypothesis of homology; (c) the 12 TTSs that result from the homology (ABCD)EF.

distribution of the homologue relative to (ABCDEF) and results in a hypothesis of relationship (homology) between those taxa (ABCD) which have the homologue and those taxa (EF) which lack the homologue. Most data sets include data which are incongruent. Any given hypothesis of homology, in the face of incongruence, will then be interpreted relative to parsimony, transformation of character states, a given unrooted network and cladistic parsimony programs' optimization criteria. TTS analysis adopts a different perspective. A hypothesis of homology can be divided prior to analysis into its smallest constituent units which are TTSs. For example, the hypothesis of homology in Fig. 8.21b can be broken down into 12 TTSs (Fig. 8.21c). In the face of incongruence none, some or all TTSs from a given hypothesis of homology may fit a tree as evidence of relationship. Any given statement coded (11)0 either fits a tree or does not, i.e. is either evidence of relationship or not. In contrast, standard parsimony analysis interprets all data within the context of plesiomorphy, homoplasy and apomorphy which is a simple model of character evolution.

Three-taxon statement analysis

TTS analysis seeks to represent propositions of homology in the form of TTSs (Nelson and Platnick, 1991; Kitching *et al.*, 1998). As noted above, homology is a proposition between taxa that share a homologue and taxa that lack that

Table 8.9 Table 8.3 re-coded as paired homologues with a one (1) representing the presence of a homologue and a zero (0) representing the absence of the homologue

Taxon	Paired homologues							Complement relations			
	1 ab	*2* ab	*3* ab	*4* ab	*5* ab	*6* ab	*7* ab	*8*	*9*	*10*	*11*
A	01	01	01	01	01	01	01	1	1	1	1
B	01	01	01	10	10	10	01	1	1	1	0
C	10	10	10	01	01	01	10	0	0	0	1
D	10	10	10	10	10	10	01	0	1	1	0
E	10	10	10	01	01	01	10	0	0	0	1

Table 8.10 Table 8.9 coded as TTSs with statements in bold and marked (*) that are accommodated on Fig. 8.22

	1a	*1b*	*2a*	*2b*	*3a*	*3b*	*4a*	*4b*	*5a*
	000000	000	000000	000	000000	000	000	000000	000
A	0?0?0?	111	0?0?0?	111	0?0?0?	111	0??	1111??	0??
B	?0?0?0	111	?0?0?0	111	?0?0?0	111	111	0?0?0?	111
C	??1111	0??	??1111	0??	??1111	0??	?0?	11??11	?0?
D	11??11	?0?	11??11	?0?	11??11	?0?	111	?0?0?0	111
E	1111??	??0	1111??	??0	1111??	??0	??0	??1111	??0
	**	***	**	***	**	***	**	**	**

	5b	*6a*	*6b*	*7a*	*7b*	*8*	*9*	*10*	*11*
	000000	000	000000	000	000000	000	000000	000000	000000
A	1111??	0??	1111??	??0	11??11	111	11??11	11??11	11??11
B	0?0?0?	111	0?0?0?	?0?	1111??	111	1111??	1111??	0?0?0?
C	11??11	?0?	11??11	111	0?0?0?	0??	0?0?0?	0?0?0?	1111??
D	?0?0?0	111	?0?0?0	0??	??1111	?0?	??1111	??1111	?0?0?0
E	??1111	??0	??1111	111	?0?0?0	??0	?0?0?0	?0?0?0	??1111
	**	**	**	***	******	***	******	******	**

Table 8.11 Number of statements (No.St.) and individual weights for each statement (I.W.) for all homologues

Homologue	No.St.	I.W.	Homologue	No.St.	I.W.
1a	6	0.66	5b	6	0.66
1b	3	1	6a	3	1
2a	6	0.66	6b	6	0.66
2b	3	1	7a	3	1
3a	6	0.66	7b	6	0.66
3b	3	1	8	3	1
4a	3	1	9	6	0.66
4b	6	0.66	10	6	0.66
5a	3	1	11	6	0.66

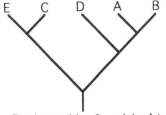

Rooting position for minimal tree

Figure 8.22 Minimal tree for TTS matrix from Table 8.10. The tree accommodates 53 TTSs with a total accommodated weight of 42.33.

homologue. Any given homology proposition can be accounted for in terms of TTSs. Kitching *et al.* (1998) provide an excellent introduction to TTS analysis within the context of Nelson and Platnick (1991). Scotland (1999) explored TTS in the context of shared homologue data but did not deal with multistate characters. Here I provide worked examples using TTS analysis for the binary data from Tables 8.1, 8.3 and 8.7 and for the mtDNA sequence data of Brown *et al.* (1982) (Appendix 8.1). All analyses were performed using Hennig86 (Farris, 1988) except for the sequence data from Appendix 8.1 which were analysed using Paup version 3.1.1 (Swofford, 1993). Fractional weighting is applied to a given homology which contains logically dependent TTSs (Nelson and Ladiges, 1992). For instance, a homology distributed (ABC)D specifies the three TTSs (AB)D, (AC)D and (BC)D, any two of which logically imply the third. Fractional weighting corrects for this logical dependency. Fractional weights were multiplied equally to whole numbers. To find the minimal tree for any TTS matrix an all-zero outgroup is added to enable the matrix to be analysed using conventional parsimony programs.

TTS analysis and Tables 8.3 and 8.7

Table 8.3

Table 8.3 includes seven paired homologues and four complement relation homologues. For the paired homologues each zero (0) and (1) entry represents data. To convert this into a matrix which can be readily understood in terms of a TTS matrix in which only the one (1) entries are informative, each column in Table 8.3 is duplicated as in Table 8.9. To convert Table 8.9 into a TTS matrix (Table 8.10), each column of one (1) entries is converted into TTSs. Table 8.10 has a total of 84 TTSs. The fractional weights for each individual statement are listed in Table 8.11.

Analysis of Table 8.10 using Hennig86 (Farris, 1988) results in one minimal tree (Fig. 8.22) in which a total of 53 TTSs are accommodated (fit) on the tree. Accommodated statements are shown in bold and marked (*) in Table 8.10. The tree (Fig. 8.22) accommodates 53 statements with a combined accommodated weight of 42.33.

Table 8.12 Table 8.7 re-coded as paired homologues with a one (1) representing the presence of a homologue and a zero (0) representing the absence of the homologue

Taxon	Paired homologues		Complement homologues
	1	2	3
	ab	ab	
A	01	01	1
B	01	01	1
C	10	01	1
D	10	10	0
E	10	10	0

Table 8.13 Table 8.12 coded as TTSs with statements in bold and marked (*) that are accommodated on Fig. 8.23

Taxa	1a	1b	2a	2b	3
OG	000000	000	000	000000	000000
A	?0?0?0	111	??0	1111??	11??11
B	0?0?0?	111	?0?	11??11	1111??
C	??1111	0??	0??	??1111	??1111
D	1111??	?0?	111	0?0?0?	0?0?0?
E	11??11	??0	111	?0?0?0	?0?0?0
	**	***	***	******	******

Table 8.14 Number of statements (No.St.) and individual weights for each statement (I.W.) for all homologues

Homologue	No.St.	I.W.
1a	6	0.66
1b	3	1
2a	3	1
2b	6	0.66
3	6	0.66

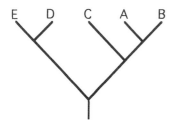

Rooting position for minimal tree

Figure 8.23 Minimal tree for TTS matrix from Table 8.13. The tree accommodates 20 TTSs with a total accommodated weight of 15.33.

Table 8.15 Data from Table 8.1 re-coded as paired homologues with a one (1) representing the presence of a homologue and a zero (0) representing the absence of the homologue. Shown for multistate data including outgroup taxa (AB) and excluding outgroup taxa

Taxon	1 abc	2 abc	3 abc	45			1 ab	2 ab	3 ab
A	100	100	100	00					
B	100	100	100	11					
C	001	001	001	11		C	01	01	01
D	001	001	010	11		D	01	01	10
E	010	010	001	11		E	10	10	01
F	010	010	010	11		F	10	10	10
G	001	001	010	11		G	01	01	10
H	001	010	010	11		H	01	10	10

Table 8.16 TTS matrix for all taxa from data in Table 8.15

```
      1a      1b      1c                                      2a      2b
OG  000000 000000 000000000000000000000000      000000 00000000000000
A   111111 0????? 0???0???0???0???0???0???      111111 0????0????0????
B   111111 ?0???? ?0???0???0???0???0???0??      111111 ?0????0????0???
C   0????? ??0??? 1111??????????????11111111    0????? ??0????0????0??
D   ?0???? ???0?? 111111111111??????????????    ?0???? ???0????0????0?
E   ??0??? 111111 ??0???0???0???0???0???0?      ??0??? 1111111111?????
F   ???0?? 111111 ??0???0???0???0???0???0?      ???0?? 1111?????11111
G   ????0? ????0? ????1111????11111111????      ????0? ????0???0????0
H   ?????0 ?????0 ????????11111111????1111      ?????0 ?????1111111111
```

```
      2c               3a      3b                              3c
OG  000000000000000      000000 0000000000000000000000000 000000
A   0????0????0????      111111 0???0???0???0???0???0??? 0?????
B   ?0????0????0???      111111 ?0???0???0???0???0???0?? ?0????
C   1111111111?????      0????? ??0??0???0???0???0???0? 111111
D   11111?????11111      ?0???? 111111111111??????????? ??0???
E   ??0????0????0??      ??0??? ??0???0???0???0???0???0 111111
F   ???0???0????0?      ???0?? 1111??????????11111111 ???0??
G   ?????1111111111      ????0? ????1111????11111111???? ????0?
H   ????0????0????0      ?????0 ????????11111111????1111 ?????0
```

```
      4                                    5
OG  000000000000000000000000000000      000000000000000000000000000000
A   0?0?0?0?0?0?0?0?0?0?0?0?0?0?0?0?      0?0?0?0?0?0?0?0?0?0?0?0?0?0?0?
B   ?0?0?0?0?0?0?0?0?0?0?0?0?0?0?0?0      ?0?0?0?0?0?0?0?0?0?0?0?0?0?0?0
C   ????????11??????11????11??1111      ????????11?????11????11??1111
D   ?????11?????11????11??11??11       ??????11??????11????11??11??11
E   ????11??????11????11????1111??      ????11??????11????11????1111??
F   ??11??????11??????11111??????      ??11??????11??????111111??????
G   11????????11111111??????????      11????????11111111??????????
H   1111111111??????????????????      1111111111??????????????????
```

Table 8.7

Table 8.7 shows two paired homologues and one complement relation homology for five taxa. These data are coded in Table 8.12 with each homologue coded one (1) and the absence of the homologue coded zero (0). The TTS matrix is shown in Table 8.13 with a total of 24 TTSs. Table 8.14 shows the number of statements for each homologue and the fractional weight of each individual statement.

Table 8.17 Number of statements and individual weights for each statement shown for all homologues from Table 8.16

Homologue	Number of statements	Weight of each statement
1a	6	1
1b	6	1
1c	24	0.5
2a	6	1
2b	15	0.66
2c	15	0.66
3a	6	1
3b	24	0.5
3c	6	1
4	30	0.33
5	30	0.33

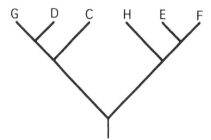

Rooting position for minimal tree

Figure 8.24 Minimal tree for TTS matrices from Tables 8.16 and 8.18. The tree accommodates 133 TTSs with a total accommodated weight of 73.5 from Table 8.16. The tree accommodates 31 TTSs with a total accommodated weight of 20.5 from Table 8.18.

Analysis of Table 8.13 using Hennig86 (Farris, 1988) results in one minimal tree (Fig. 8.23) in which a total of 20 TTSs are accommodated (fit) on the tree. Accommodated statements are shown in bold and marked (*) in Table 8.13. The tree (Fig. 8.23) accommodates 20 statements with a combined accommodated weight of 15.33.

Fig. 8.23 can be compared to the standard analysis of these data shown in Figs. 8.17b–d. The result from the TTS analysis (Fig. 8.23) agrees with Fig. 8.17d from the standard analysis but excludes Fig. 8.17c.

TTS analysis and Table 8.1

The multistate coding from Table 8.1 can be converted into binary data with a one (1) representing the presence of each homologue and a zero (0) representing the absence of each homologue as in Table 8.15. The TTS matrix results in 168 TTSs and is shown in Table 8.16. The fractional weights for each statement are given in Table 8.17. Analysis of Table 8.16 results in a single minimal tree (Fig. 8.24, outgroups not shown) with 133 accommodated TTSs with a total accommodated weight of 73.5. Similarly, the binary data excluding outgroup taxa from Table 8.1 can be converted into a TTS matrix (Table 8.18) which in this case

Table 8.18 TTS matrix for ingroup taxa from data in Table 8.15

	1a	1b	2a	2b	3a	3b
Og	0000	000000000000	000000000	000000000	000000000000	0000
C	0???	111111??????	0??0??0??	111111???	0?0?0?0?0?0?	1111
D	?0??	11????11111??	?0??0??0?	111???111	????11??1111	0???
E	1111	0?0?0?0?0?0?	111111???	0??0??0??	?0?0?0?0?0?0	1111
F	1111	?0?0?0?0?0?0	111???111	?0??0??0?	??11??11??11	?0??
G	??0?	??11??11??11	??0??0??0	???111111	11????11111??	??0?
H	??0?	????11??1111	???111111	??0??0??0	111111??????	??0?

Table 8.19 Number of statements and individual weights for each statement shown for all homologues from Table 8.18

Homologue	Number of statements	Weight of each statement
1a	4	1
1b	12	0.5
2a	9	0.66
2b	9	0.66
3a	12	0.5
3b	4	1

excludes taxa A and B, and this renders the data binary for all character states. The fractional weights for each statement are given in Table 8.19. Analysis of Table 8.18 results in a single minimal tree (Fig. 8.24) which accommodates 31 statements with a total combined accommodated weight of 20.5. Both TTS matrices (Tables 8.16 and 8.18) yield the same tree for these data (Fig. 8.24). This is perhaps not surprising given that taxa A and B have no character states in common with the ingroup taxa. However, as shown earlier in this chapter, these data treated as multistate and binary data as in Table 8.1 led to completely different results, indicating a discrepancy between binary and multistate representation of the same data.

TTS analysis and sequence data

Appendix 8.1 represents five mitochondrial DNA sequences published by Brown *et al.* (1982) from human, chimp, gorilla, orang-utan and gibbon. The sequences are 896 bp in length and contain the genes for three transfer RNAs and parts of two proteins. Standard parsimony analysis of these sequences yields a single unrooted tree of length 145 (Fig. 8.25a) as reported in Brown *et al.* (1982). The three shortest trees are shown here as Figs 8.25a-c. The trees are shown as rooted between man, chimp, gorilla and the other two taxa.

The sequences (Appendix 8.1) have a total of 281 variable positions. With n taxa, there are $2^n - (n + 2)$ possible types of cladistically informative characters; with five taxa there are 25 such types – ten grouping two taxa, ten grouping three taxa and five grouping four taxa (Patterson, 1988b). In the anthropoid mtDNA sequences all 25 occur (Patterson, 1988b). With C = *Pan*, G = *Gorilla*, H = *Hylobates*, M = *Homo*, O = *Pongo*, the number of occurrences of each type is

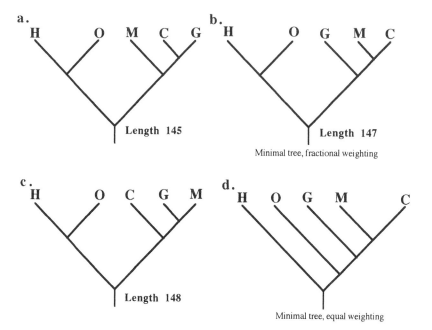

Figure 8.25 Three shortest trees for the data set of Brown *et al.* (1982) from Appendix 8.1 analysed with Hennig86 and conventional parsimony analysis. Length is shown on each tree. (b) Minimal tree for fractionally weighted TTS matrix; (d) minimal tree for equally weighted TTS matrix.

shown in Table 8.20. Table 8.20 shows the number of TTSs that result from these data, the total number of statements, the individual weight of each statement and the total weight of all statements. The TTS matrix (not shown) contains a total of 1926 TTSs.

Analysis of the TTS matrix for these data yields a single minimal tree (Fig. 8.25b). This tree accommodates a total of 905 TTSs with a total accommodated weight of 591.66. The next two minimal trees in the TTS analysis are Figs 8.25a and c. The number of accommodated TTSs and accommodated weights are 886 accommodated TTSs and total accommodated weight 579.83 (Fig. 8.25a), and 890 accommodated TTSs and total accommodated weight 580.33 (Fig. 8.25c).

The minimal tree (Fig. 8.25b) from the TTS analysis, which pairs man with chimp, agrees with the results of Hasegawa and Yano (1984) and Bishop and Friday (1986) for these data using maximum likelihood. The same tree was found by Patterson (1988b) using an eclectic form of compatibility analysis. Nei *et al.* (1985) found the same tree from a UPGMA analysis of these data. There are 19 more TTSs accommodated on the (man+chimp) tree than the (gorilla+chimp) tree, which is the most parsimonious result using standard cladistic analysis. The results from the TTS analysis are in agreement with those of several other studies but at odds with the results of standard parsimony analysis. Analysis of these data without fractional weighting results is shown in Fig. 8.25d.

Table 8.20 Sequence data of Brown *et al.* (1982) showing patterns of shared bases in pairs, threes and fours. Number of statements, running total of statements, individual weight of each statement and total weight of statements are also given

Taxa	Shared pairs	Number of statements	Running total of statements	Individual weight	Total weight
CG	12	36	36	1	36
CH	6	18	54	1	18
CM	14	42	96	1	42
CO	4	12	108	1	12
GH	8	24	132	1	24
GM	9	27	159	1	27
GO	7	21	180	1	21
HM	5	15	195	1	15
HO	32	96	291	1	96
MO	3	9	300	1	9
Shared threes					
CGH	3	18	318	0.66	12
CGM	31	186	504	0.66	124
CGO	7	42	546	0.66	28
CHM	7	42	588	0.66	28
CHO	5	30	618	0.66	20
CMO	11	66	684	0.66	44
GHM	3	18	702	0.66	12
GHO	10	60	762	0.66	40
GMO	10	60	822	0.66	40
HMO	10	60	882	0.66	40
Shared fours					
CGHM	52	312	1194	0.5	156
CGHO	15	90	1284	0.5	45
CGMO	64	384	1668	0.5	192
CHMO	24	144	1812	0.5	72
GHMO	19	114	1926	0.5	57

A question concerning the TTS analyses which result in Fig. 8.25b (fractional weighting) or Fig. 8.25d (equal weighting) is whether each tree topology is rooted or not and the taxonomic implications for included taxa. In any TTS analysis the all zero outgroup roots the tree in the position that maximizes fit between accommodated statements, accommodated weight and a given solution. The solution is therefore rooted. However, the solution has different meaning for ingroup and outgroup taxa. It has been shown that for any systematic problem it is advisable to include related (outgroup) taxa which in effect are placeholders for the tree of life. The ingroup taxa attach to the outgroup taxa in such a way that a globally most parsimonious solution is found. Any analysis of ingroup taxa alone is open to the accusation of being a locally, not globally, parsimonious solution (Weston, 1994). Outgroup taxa are therefore included to test ingroup monophyly and facilitate a globally most parsimonious solution. For example, if the relationships were sought between the five taxa discussed above then it would be necessary to include related

Table 8.21 Analogy, homology and non-homology can be distinguished relative to similarity and the concept of fit. Analogy fails the similarity test and should not be coded in a matrix. Homology passes the similarity test (topological correspondence) and in the form of TTSs fits a node on a tree. Non-homology is topological similarity that does not provide evidence of relationship. Non-homology in this sense is directly equivalent to Lankaster's (1870) use of the term 'homoplasy', i.e. topological correspondence that is not evidence of relationship

	Analogy	Homology	Non-homology
Similarity	Fail	PASS	PASS
Fit	N/A	EXACT	FAIL

taxa to test ingroup monophyly and estimate a global rather than local solution to the problem. For the analysis presented above, there is good reason from the results presented in previous studies to suspect that man, chimp and gorilla are a monophyletic group relative to the rest of life. This hypothesis can be tested in the context of related taxa, in this case gibbon and orang-utan. Figs 8.25b and d, the results of the TTS analyses, show that ingroup monophyly is established and that a globally parsimonious solution for this has been estimated in the context of related taxa. The solutions for the outgroup taxa are local solutions because if these were the taxa of interest then a sample of related taxa would need to be larger than the five taxa sampled here to take into account the putative paraphyly of orang-utan and gibbon relative to man, chimp and gorilla and Old World monkeys (Andrews, 1987).

Patterson's tests of homology and TTS analysis

Patterson's (1982) similarity and congruence tests of homology are shown in Table 8.5. Table 8.6 shows these tests reinterpreted in the context of standard cladistic analysis relative to similarity and fit. Table 8.21 shows that for TTS analysis any set of statements either fit (are accommodated on a tree) or do not fit (are not accommodated on a tree) a tree. TTS analysis serves only to test whether hypotheses of homology in the form of TTSs are homologous (evidence of relationship) or are non-homologous (not evidence of relationship).

Conclusions

I set out to explore the issue of whether the data presented in Table 8.1 contain a cladistic structure for resolving relationships between taxa (CDEFGH). Fig. 8.24 demonstrates that analysis of these data using the TTS approach results in a rooted solution for these data. These data analysed using multistate coding result in an unresolved bush (Fig. 8.1), or with binary coding result in the unrooted tree (Fig. 8.2a). The TTS approach treats all character states as separate hypotheses of homology, whereas the standard approach treats them as part of a transformation series. From a taxic perspective it is very difficult to think of a conditional phrase, for any of the character states of the first three characters. Even assuming that taxa A and B are outgroup taxa, in what sense can 2+4 stamens, red+blue flowers, palmate+simple leaves constitute homology propositions that the states 5 stamens, brown flowers and pinnate leaves do not belong? If the homology propositions involved all three character states of stamen number, flower colour and leaf type, then each homology would be uninformative as all eight taxa would be included. Therefore, for each of the first three characters in Table 8.1 these data seem to lead to hypotheses of taxic homology at the level of 2 stamens, 4 stamens, 5 stamens, brown flowers, blue flowers, red flowers, pinnate leaves, palmate leaves and simple leaves. To code all three character states, for the first three characters, as a transformation series, is to assume the monophyletic origin of all three structures (the column in a standard cladistic data matrix is equivalent to a statement of plesiomorphy, and if plesiomorphy is apomorphy at another hierarchical level then the homology is assumed never tested). Standard cladistic analysis tests how many times a transformation occurs but never questions the transformations. As shown by Julie Hawkins (Chapter 2), many of the problems with character coding in the current cladistic literature relate to transformational hypotheses of homology. In other words, there are as many hypotheses of *what if* transformations as there are systematic botanists. A taxic view of homology does not solve all these problems, but it provides a focus in that, at some comparative level, different structures that are being hypothesized as homologous should be the same relative to other taxa which lack the homology. For example, Crane *et al.* (1995) list homologies for angiosperms, relative to other seed plants, including: sieve tube and companion cells from same initials, stamens with two pairs of pollen sacs, anthers with hypodermal endothecium, microgametophyte of only three nuclei, carpel with stigmatic pollen germination and triploid endosperm. These homologies characterize the taxon angiosperms relative to other seed plants. In contrast, the homologies for characters 1–3 from Table 8.1 are at the level of leaf type, stamen number and flower colour. These three homologies characterize all taxa and are therefore uninformative as they led to no new hypotheses of relationship. These types of homologies are what Owen termed 'general homologies' (Patterson, 1982: 35). Patterson (1982: 35) went on to distinguish archetypes from morphotypes: 'An archetype is an idealization with which features of organisms may be homologized by abstract transformations which entail no hypotheses of hierarchic grouping' In this quote, replace 'archetype' with 'standard cladistic data matrix' to appreciate that archetypes *sensu* Patterson (1982) are by no means extinct.

Patterson's (1982) taxic view of homology and Nelson's (1989a) view of homology as relationship represent interpretations of homology which are grounded in the idea that homology is a relational concept of hierarchical grouping. Standard cladistic data matrices contain general transformational homologies which lead to no new hypotheses of grouping independent of the root of the tree. TTS analysis represents a method of character coding in which Nelson and Platnick's (1991) interpretation of taxic homology can be explored and implemented.

Acknowledgements

I thank Jonathan Bennett, Mark Carine and Elizabeth Moylan for comments and putting up with me during the summer of 1998. Thanks also to Gary Nelson and Dave Williams for many fruitful discussions and to Chris Humphries and Dave Williams for comments on the manuscript. I am funded by The Royal Society.

Appendix 8.1: mtDNA sequence data of Brown et al. (1982)

```
                       1111111111222222222223333333333344444444445555555555566666
                       1234567890123456789012345678901234567890123456789012345678901234
```

Human	AAGCTTCACCGGCGCAGTCATTCTCATAATCGCCCACGGACTTACATCCTCATTACTATTCTGC
Chimpanzee	AAGCTTCACCGGCGCAATTATCCTCATAATCGCCCACGGACTTACATCCTCATTATTATTCTGC
Gorilla	AAGCTTCACCGGCGCAGTTGTTCTTATAATTGCCCACGGACTTACATCATCATTATTATTCTGC
Orang-utan	AAGCTTCACCGGCGCAACCACCCTCATGATTGCCCATGGACTCACATCCTCCCTACTGTTCTGC
Gibbon	AAGCTTTACAGGTGCAACCGTCCTCATAATCGCCCACGGACTAACCTCTTCCCTGCTATTCTGC

```
                                       11111111111111111111111111111
                       6666677777777777888888888889999999999900000000000011111111112222222222
                       5678901234567890123456789012345678901234567890123456789012345678
```

Human	CTAGCAAACTCAAACTACGAACGCACTCACAGTCGCATCATAATCCTCTCTCAAGGACTTCAAA
Chimpanzee	CTAGCAAACTCAAATTATGAACGCACCCACAGTCGCATCATAATTCTCTCCCAAGGACTTCAAA
Gorilla	CTAGCAAACTCAAACTACGAACGAACCCACAGCCGCATCATAATTCTCTCTCAAGGACTCCAAA
Orang-utan	CTAGCAAACTCAAACTACGAACGAACCCACAGCCGCATCATAATCCTCTCTCAAGGCCTTCAAA
Gibbon	CTTGCAAACTCAAACTACGAACGAACTCACAGCCGCATCATAATCCTATCTCGAGGGCTCCAAG

```
                       1111111111111111111111111111111111111111111111111111111111111111
                       2333333333344444444445555555555566666666666777777777788888888888999
                       9012345678901234567890123456789012345678901234567890123456789012
```

Human	CTCTACTCCCACTAATAGCTTTTTGATGACTTCTAGCAAGCCTCGCTAACCTCGCCTTACCCCC
Chimpanzee	CTCTACTCCCACTAATAGCCTTTTGATGACTCCTAGCAAGCCTCGCTAACCTCGCCCTACCCCC
Gorilla	CCCTACTCCCACTAATAGCCCTTTGATGACTTCTGGCAAGCCTCGCCAACCTCGCCTTACCCCC
Orang-utan	CTCTACTCCCCCTAATAGCCCTCTGATGACTTCTAGCAAGCCTCACTAACCTTGCCCTACCACC
Gibbon	CCTTACTCCCACTGATAGCCTTCTGATGACTCGCAGCAAGCCTCGCTAACCTCGCCCTACCCCC

```
                       1111111222222222222222222222222222222222222222222222222222222222
                       9999999900000000001111111111222222222233333333334444444444555555555
                       3456789012345678901234567890123456789012345678901234567890123456
```

Human	CACTATTAACCTACTGGGAGAACTCTCTGTGCTAGTAACCACGTTCTCCTGATCAAATATCACT
Chimpanzee	TACCATTAATCTCCTAGGGGAACTCTCCGTGCTAGTAACCTCATTCTCCTGATCAAATACCACT
Gorilla	CACCATTAACCTACTAGGAGAGCTCTCCGTACTAGTAACCACATTCTCCTGATCAAACACCACC
Orang-utan	CACCATCAACCTTCTAGGAGAACTCTCCGTACTAATAGCCATATTCTCTTGATCAACATCACC
Gibbon	CACTATTAACCTCCTAGGTGAACTCTTCGTACTAATGGCCTCCTTCTCCTGGCAAACACTACT

```
                       2222222222222222222222222222222222222222222333333333333333333333
                       5556666666666777777777788888888889999999999000000000001111111111 2
                       7890123456789012345678901234567890123456789012345678901234567890
```

Human	CTCCTACTTACAGGACTCAACATACTAGTCACAGCCCTATACTCCCTCTACATATTTACCACAA
Chimpanzee	CTCCTACTCACAGGATTCAACATACTAATCACAGCCCTGTACTCCCTCTACATGTTTACCACAA
Gorilla	CTTTTACTTACAGGATCTAACATACTAATTACAGCCCTGTACTCCCTTTATATATTTACCACAA
Orang-utan	ATCCTACTAACAGGACTCAACATACTAATCACAACCCTATACTCTCTCTATATATTCACCACAA
Gibbon	ATTACACTCACCGGGCTCAACGTACTAATCACGGCCCTATACTCCCTTTACATATTTATCATAA
```
```

```
            33333333333333333333333333333333333333333333333333333333333333
            22222222233333333334444444444455555555556666666666777777777788888
            12345678901234567890123456789012345678901234567890123456789012 34
```

Human   CACAATGGGGCTCACTCACCCACCACATTAACAACATAAAACCCTCATTCACACGAGAAAACAC
Chimpanzee CACAATGAGGCTCACTCACCCACCACATTAATAACATAAAGCCCTCATTCACACGAGAAAATAC
Gorilla   CACAATGAGGCCCACTCACACACCACATCACCAACATAAAACCCTCATTTACACGAGAAAACAT
Orang-utan CACAACGAGGTACACCCACACACCACATCAACAACATAAAACCTTCTTTCACACGCGAAAATAC
Gibbon   CACAACGAGGCACACTTACACACCACATTAAAAACATAAAACCCTCACTCACACGAGAAAACAT

```
            33333333333333334444444444444444444444444444444444444444444444
            88888999999999900000000000111111111122222222223333333333444444444
            5678901234567890123456789012345678901234567890123456789012345678
```

Human   CCTCATGTTCATACACCTATCCCCCATTCTCCTCCTATCCCTCAACCCCGACATCATTACCGGG
Chimpanzee TCTCATATTTTTACACCTATCCCCCATCCTCCTTCTATCCCTCAATCCTGATATCATCACTGGA
Gorilla   CCTCATATTCATGCACCTATCCCCCATCCTCCTCCTATCCCTCAACCCCGATATTATCACCGGG
Orang-utan CCTCATGCTCATACACCTATCCCCCATCCTCCTCTTATCCCTCAACCCCAGCATCATCGCTGGG
Gibbon   ATTAATACTTATGCACCTCTTCCCCCTCCTCCTCCTAACCCTCAACCCTAACATCATTACTGGC

```
            44444444444444444444444444444444444444444444444444445555555555555
            45555555555566666666667777777777788888888889999999999900000000000111
            90123456789012345678901234567890123456789012345678901234 56789012
```

Human   TTTTCCTCTTGTAAATATAGTTTAACCAAAACATCAGATTGTGAATCTGACAACAGAGGCTTAC
Chimpanzee TTCACCTCCTGTAAATATAGTTTAACCAAAACATCAGATTGTGAATCTGACAACAGAGGCTCAC
Gorilla   TTCACCTCCTGTAAATATAGTTTAACCAAAACATCAGATTGTGAATCTGATAACAGAGGCTCAC
Orang-utan TTCGCCTACTGTAAATATAGTTTAACCAAAACATTAGATTGTGAATCTAATAATAGGGCCCCAC
Gibbon   TTTACTCCCTGTAAACATAGTTTAATCAAAACATTAGATTGTGAATCTAACAATAGAGGCTCGA

```
            55555555555555555555555555555555555555555555555555555555555555
            11111112222222222333333333344444444445555555555566666666667777777
            34567890123456789012345678901234567890123456789012345678901 23456
```

Human   GACCCCTTATTTACCGAGAAAGCTCACAAGAACTGCTAACTCATGCCCCCATGTCTGACAACAT
Chimpanzee GACCCCTTATTTACCGAGAAAGCTTATAAGAACTGCTAATTCATATCCCCATGCCTGACAACAT
Gorilla   AACCCCTTATTTACCGAGAAAGCTCGTAAGAGCTGCTAACTCATACCCCCGTGCTTGACAACAT
Orang-utan AACCCCTTATTTACCGAGAAAGCTCACAAGAACTGCTAACTCTCACT?CCATGTGTGACAACAT
Gibbon   AACCTCTTGCTTACCGAGAAAGCCCACAAGAACTGCTAACTCACTATCCCATGTATGACAACAT

```
            55555555555555555555555556666666666666666666666666666666666666
            77788888888888999999999900000000000111111111122222222223333333333 4
            78901234567890123456789012345678901234567890123456789012345 67890
```

Human   GGCTTTCTCAACTTTTAAAGGATAACAGCTATCCATTGGTCTTAGGCCCCAAAAATTTTGGTGC
Chimpanzee GGCTTTCTCAACTTTTAAAGGATAACAGCCATCCGTTGGTCTTAGGCCCCAAAAATTTTGGTGC
Gorilla   GGCTTTCTCAACTTTTAAAGGATAACAGCTATCCATTGGTCTTAGGACCCAAAAATTTTGGTGC
Orang-utan GGCTTTCTCAGCTTTTAAAGGATAACAGCTATCCCTTGGTCTTAGGATCCAAAAATTTTGGTGC
Gibbon   GGCTTTCTCAACTTTTAAAGGATAACAGCTATCCATTGGTCTTAGGACCCAAAAATTTTGGTGC

```
                    6666666666666666666666666666666666666666666666666666666666677777
                    4444444445555555555566666666666777777777788888888889999999999900000
                    1234567890123456789012345678901234567890123456789012345678901234
```

Human   AACTCCAAATAAAAGTAATAACCATGCACACTACTATAACCACCCTAACCCTGACTTCCCTAAT
Chimpanzee   AACTCCAAATAAAAGTAATAACCATGTATACTACCATAACCACCTTAACCCTAACTCCCTTAAT
Gorilla   AACTCCAAATAAAAGTAATAACTATGTACGCTACCATAACTGCCTTAGCCCTAACTTCCTTAAT
Orang-utan   AACTCCAAATAAAAGTAACAGCCATGTTTACCACCATAACCACCCTCACCTTAACTTCCCTAAT
Gibbon   AACTCCAAATAAAAGTAATAGCAATGTACACCACCATAGCCATTCTAACGCTAACCTCCCTAAT

```
                    7777777777777777777777777777777777777777777777777777777777777777
                    0000011111111112222222222233333333333444444444455555555555666666666
                    5678901234567890123456789012345678901234567890123456789012345678
```

Human   TCCCCCCATCCTTACCACCCTCGTTAACCCTAACAAAAAAAACTCATACCCCCATTATGTAAAA
Chimpanzee   TCTCCCCATCCTCACCACCCTCATTAACCCTAACAAAAAAAACTCATATCCCCATTATGTGAAA
Gorilla   TCCCCCTATCCTTACCACCTTCATCAATCCTAACAAAAAAGCTCATACCCCCATTACGTAAAA
Orang-utan   CCCCCCCATTACCGCTACCCTCATTAACCCCAACAAAAAAAACCCATACCCCCACTATGTAAAA
Gibbon   TCCCCCCATTACAGCCACCCTTATTAACCCCAATAAAAAGAACTTATACCCGCACTACGTAAAA

```
                    7777777777777777777777777777777888888888888888888888888888888888
                    6777777777788888888889999999999900000000000111111111122222222223333
                    9012345678901234567890123456789012345678901234567890123456789012
```

Human   TCCATTGTCGCATCCACCTTTATTATCAGTCTCTTCCCCACAACAATATTCATGTGCCTAGACC
Chimpanzee   TCCATTATCGCGTCCACCTTTATCATTAGCCTTTTCCCCACAACAATATTCATATGCCTAGACC
Gorilla   TCTATCGTCGCATCCACCTTTATCATCAGCCTCTTCCCCACAACAATATTTCTATGCCTAGACC
Orang-utan   ACGGCCATCGCATCCGCCTTTACTATCAGCCTTATCCCAACAACAATATTTATCTGCCTAGGAC
Gibbon   ATGACCATTGCCTCTACCTTTATAATCAGCCTATTTCCCACAATAATATTCATGTGCACAGACC

```
                    8888888888888888888888888888888888888888888888888888888888888888
                    3333333444444444455555555556666666666777777777788888888889999999
                    3456789012345678901234567890123456789012345678901234567890123456
```

Human   AAGAAGTTATTATCTCGAACTGACACTGAGCCACAACCCAAACAACCCAGCTCTCCCTAAGCTT
Chimpanzee   AAGAAGCTATTATCTCAAACTGGCACTGAGCAACAACCCAAACAACCCAGCTCTCCCTAAGCTT
Gorilla   AAGAAGCTATTATCTCAAGCTGACACTGAGCAACAACCCAAACAATTCAACTCTCCCTAAGCTT
Orang-utan   AAGAAACCATCGTCACAAACTGATGCTGAACAACCACCCAGACACTACAACTCTCACTAAGCTT
Gibbon   AAGAAACCATTATTTCAAACTGACACTGAACTGCAACCCAAACGCTAGAACTCTCCCTAAGCTT

References

Andrews, P. (1987) Aspects of homonoid phylogeny, in Patterson, C. (ed.) *Molecules and Morphology: Conflict or Compromise*, Cambridge: Cambridge University Press, pp. 23–53.

Bishop, M.J. and Friday, A.E. (1986) Molecular sequences and hominoid phylogeny, in Wood, B., Martin, L. and Andrews, P.J. (eds) *Major Topics in Primate and Human Evolution*, Cambridge: Cambridge University Press, pp. 150–156.

Brady, R. (1985) On the independence of systematics, *Cladistics*, 1, 113–126.

Brown, W.M., Praeger, E.M., Wang, A. and Wilson, A.C. (1982) Mitochondrial DNA sequences of primates: tempo and mode of evolution, *Journal of Molecular Evolution*, 18, 225–239.

Carine, M.A. and Scotland, R.W. (1999) Taxic and transformational homology: different ways of seeing, *Cladistics*.

Crane, P.R., Friis, E.M. and Pederson, K.R. (1995) The origin and early diversification of angiosperms, *Nature*, **374**, 27–33.

Dawkins, R. (1986) *The Blind Watchmaker*, London: Penguin Books.

De Pinna, M.C.C. (1991) Concepts and tests of homology in the cladistic paradigm, *Cladistics*, **7**, 317–338

Eldredge, N. (1979) Alternative approaches to evolutionary theory, *Bulletin of the Carnegie Museum of Natural History*, **13**, 7–19.

Farris, J.S. (1983) The logical basis of phylogenetic analysis, in Platnick, N.I. and Funk, V.A. (eds) *Advances in Cladistics*, vol. 2, New York: Columbia University Press, pp. 7–36.

Farris, J.S. (1988) *Hennig86 Version 1.5. MS-DOS Program*, Port Jefferson Station, New York: published by the author.

Farris, J.S., Kluge, A.G. and Eckhardt, M.J. (1970) A numerical approach to phylogenetic systematics, *Systematic Zoology*, **19**, 172–191.

Futuyma, D.J. (1986) *Evolutionary Biology*, Sunderland, MA: Sinauer Associates.

Hasegawa, M. and Yano, T. (1984) Phylogeny and classification of Hominoidea as inferred from DNA sequence data, *Proceedings of the Japanese Academy*, **60B**, 389–392.

Hawkins, J.A., Hughes, C.E. and Scotland, R.W. (1997) Primary homology assessment, characters and character states, *Cladistics*, **13**, 275–283.

Hennig, W. (1966) *Phylogenetic Systematics*, Urbana, IL: University of Illinois Press.

Hull, D.L. (1979) The limits of cladism, *Systematic Zoology*, **28**, 414–438.

Kenrick, P. and Crane, P.R. (1997) *The Origin and Early Evolution of Land Plants: a Cladistic Study*, Washington, DC: Smithsonian Institution Press.

Kitching, I.L. (1991) The determination of character polarity, in Forey, C.J., Humphries, C.J., Kitching, I.L., Scotland, R.W., Siebert, D.J. and Williams, D.M. (eds) *Cladistics: a Practical Course in Systematics*, Oxford: Clarendon Press, pp. 22–43.

Kitching, I.L., Forey, P.L., Humphries, C.J. and Williams, D.M. (1998) *Cladistics: The Theory and Practice of Parsimony Analysis*, Oxford: Oxford University Press.

Lankaster, E.R.(1870) On the use of the term homology in modern zoology, and the distinction between homogenetic and homoplastic agreements, *The Annals and Magazine of Natural History*, **6**, 35–43.

Meacham, C.A. and Estabrook, G.F. (1985) Compatibility methods in systematics, *Annual Review of Ecology and Systematics*, **116**, 431–446.

Nei, M., Stephens, J.C. and Saitou, N. (1985) Methods for computing the standard errors in branching points in an evolutionary tree and their application to molecular data from humans and apes, *Molecular Biology and Evolution*, **2**, 66–85.

Nelson, G.J. (1970) Outline of a theory of comparative biology, *Systematic Zoology*, **19**, 373–384.

Nelson, G.J. (1978) Ontogeny, phylogeny, palaeontology, and the biogenetic law, *Systematic Zoology*, **27**, 324–345.

Nelson, G.J. (1989a) Cladistics and evolutionary models, *Cladistics*, **5**, 275–289.

Nelson, G.J. (1989b) Species and taxa: systematics and evolution, in Otte, D. and Endler, J.A. (eds) *Speciation and its Consequences*, Sunderland, MA: Sinauer Associates, pp. 60–81.

Nelson, G.J. (1994) Homology and systematics, in Hall, B.K. (ed.) *Homology: the Hierarchical Basis of Comparative Biology*, San Diego, CA: Academic Press, pp. 101–149.

Nelson, G.J. and Ladiges, P.Y. (1992) Information content and fractional weight of three-item statements, *Systematic Biology*, **41**, 490–494.

Nelson, G.J. and Platnick, N.I. (1981) *Systematics and Biogeography: Cladistics and Vicariance*, New York: Columbia University Press.

Nelson, G.J. and Platnick, N.I. (1991) TTSs: a more precise use of parsimony? *Cladistics*, 7, 351–366.

Nixon, K.C. and Carpenter, J.M. (1993) On outgroups, *Cladistics*, 9, 413–426.

Owen, R. (1843) *Lectures on comparative anatomy*, London: Longman, Brown, Green and Longmans.

Panchen, A.L. (1994) Richard Owen and the concept of homology, in Hall, B.K. (ed.) *Homology: The Hierarchical Basis of Comparative Biology*, San Diego, CA: Academic Press, pp. 21–62.

Patterson, C. (1982). Morphological characters and homology, in Joysey, K.A. and Friday, A.E. (eds) *Problems in Phylogenetic Reconstruction*, London: Academic Press, pp. 21–74.

Patterson, C. (1988a) Homology in classical and molecular biology, *Molecular Biology and Evolution*, 5, 603–625.

Patterson, C. (1988b) The impact of evolutionary theories on systematics, in Hawksworth, D.L. (ed.) *Prospects in Systematics*, Oxford: Clarendon Press, pp. 59–91.

Platnick, N.I. (1977) Cladograms, phylogenetic trees and hypothesis testing, *Systematic Zoology*, 26, 438–442.

Platnick, N.I. (1979) Philosophy and the transformation of cladistics, *Systematic Zoology*, 28, 537–546.

Platnick, N.I. (1985) Philosophy and the transformation of cladistics revisited, *Cladistics*, 1, 87–94.

Platnick, N.I., Humphries, C.J., Nelson, G.J. and Williams, D.M. (1996) Is Farris optimization perfect? Three taxon statements and multiple branching, *Cladistics*, 12, 243–252.

Pleijel, F. (1995) On character coding for phylogeny reconstruction, *Cladistics*, 11, 309–315.

Ridley, M. (1983) Can classification do without evolution? *New Scientist*, 100, 647–651.

Ridley, M. (1996) *Evolution*, Cambridge, MA: Blackwell Science.

Ridley, M. (1997) Can classification do without evolution?, in Ridley, M. (ed.) *Evolution*, Oxford: Oxford University Press, pp. 196–206.

Riedl, R. (1982) The role of morphology in the theory of evolution, in Grene, M. (ed.) *Dimensions of Darwinism*, Cambridge: Cambridge University Press, pp. 205–238.

Rieppel, O. (1988) *Fundamentals of Comparative Biology*, Basel: Birkhauser Verlag.

Rieppel, O. (1994) Homology, topology, and typology: the history of modern debates, in Hall, B.K. (ed.) *Homology: the Hierarchical Basis of Comparative Biology*, San Diego, CA: Academic Press, pp. 63–100.

Sattler, R. (1984) Homology – a continuing challenge, *Systematic Botany*, 9, 382–394.

Scotland, R.W. (1997) Parsimony neither maximizes congruence nor minimizes incongruence or homoplasy, *Taxon*, 46, 743–746.

Scotland, R.W. (1999) Taxic homology and TTS analysis, *Systematic Biology*.

Siebert, D.J. and Williams, D.M. (1998) Recycled, *Cladistics*, 14, 339–347.

Stevens, P.F. (1980) Evolutionary polarity of character states, *Annual Review of Ecology and Systematics*, 11, 333–358.

Swofford, D.L. (1993) *Phylogenetic Analysis Using Parsimony, Version 3.1.1*, Illinois Natural History Survey, Champaign, IL.

Weston, P.H. (1994) Methods for rooting cladistic trees, in Scotland, R.W., Siebert, D.J. and Williams, D.M. (eds), *Models in Phylogeny Reconstruction*, Oxford: Clarendon Press, pp. 125–155.

Wilkinson, M. (1997) Congruence, consistency and the 'axioms' of cladistics, *Taxon*, 46, 739–742.

Chapter 9

Characters, homology and three-item analysis

David M. Williams and Darrell J. Siebert

Introduction

Systematics is commonly, but not universally, understood in terms of phylogeny reconstruction, with results presented as phylogenetic trees. Cladistics is a powerful approach to systematics. It seeks to solve the problem of the structure of the natural world, through the discovery of characters and taxa, with results presented as clado-grams, through which 'the data of biology integrate at the level of classification' (Nelson and Platnick, 1984: 154). Cladograms are not phylogenetic trees, but are related to them in a complex fashion (Nelson and Platnick, 1981: 14, 135–151, 171–172). A cladogram may summarize a suite of several, or many, phylogenetic trees and thus may be considered to be more general than a phylogenetic tree. Cladistics may then be considered to be more general than phylogeny reconstruction. Cladistics seeks to solve the problem of the structure of the natural world by addressing it in the following way: 'What are the interrelationships among a suite of taxa?' The solution to this question is conceptually simple. For any cladistic problem the number of terminal taxa is specified at the outset and there is a finite number of possible solutions when the solutions are presented as hierarchical branching diagrams (Felsenstein, 1978). For instance, for four taxa the total number of possible solutions is 26, with 15 fully resolved solutions (all inter-relationships solved), 10 partially resolved solutions (some inter-relationships solved), and one completely unresolved solution (no inter-relationships solved). Resolved solutions suggest that among the taxa being considered, some are more closely related to each other than they are to the remainder; taxa at higher levels are thus discovered. Choosing among solutions may be done on the basis of how well available data fit each one, with data originating in observations of specimens (in the case of morphology, cell structure, etc.) or of gels (in the case of molecular sequences).

It is possible, in principle at least, to inspect all solutions by fitting the data to every possible branching diagram. All but those solutions that accommodate all the data most efficiently can be excluded ('induction by complete exclusion' (Patterson, 1981: 451), or 'deduction in disguise' (Hull 1973: 22)). This procedure embodies no more than the application of the principle of parsimony – how well do data fit particular trees? While a great deal has been written about parsimony, the appli-cation of the principle need not suggest or comply with any specific kind of implementation (Siebert and Williams, 1997, 1998).

In 'normal science' systematic problems change over time, principally in two aspects: more data accumulate; more terminal taxa may become relevant. Increasing the number of taxa simply changes the nature of the problem (Platnick, 1977). A four-taxon problem (with 26 possible solutions) becomes a five-taxon problem (with 236 possible solutions) with the inclusion of one additional taxon. In general, the number of taxa included may reflect a more appropriate focus for the problem at hand, progress thus being achieved. Real progress may also be achieved by the acquisition of new data, either by the investigation of new organismal features or by a deeper understanding of those already studied. A further aspect, only recently given attention, is, in Platnick's words, ' ... how best to represent organismic variation in the matrix' (Platnick, 1989b: 21). In recent cladistic literature this discussion has taken the form of examining particular kinds of formalisms associated with conventional matrix construction (for example Pleijel, 1995). This aspect deserves further attention, as its clarification is central to the aims of this volume. We outline an approach below that seems to contribute to this clarification and to hold much promise for the future of cladistics and comparative organismal biology by moving away from some of the accepted matrix conventions.

Homology

De Pinna's (1991) treatment of homology is a useful summary from a cladistic perspective. In his review, de Pinna (1991: 372) noted that '[E]very proposition of homology involves two stages, which are associated with its generation and legitimation'. He used the term 'primary homology' for the first stage ('generation') and 'secondary homology' for the second stage ('legitimation'). Each of these stages may be considered complex and interrelated. Primary homology considers how homology statements originate (or how they come into being through the efforts of individual investigators) and how they are subsequently represented in a form amenable to analysis. Primary homology statements are recognized as conjectures that are the product of the individual effort of investigators. Secondary homology considers the analysis of a collection of primary homology statements with the aim of identifying corroborated conjectures of taxon relationships, which themselves might eventually lead to an understanding of 'general synapomorphy' (Nelson and Platnick, 1981: 163). Corroborated statements are summarized as a cladogram, where the branching aspect, '...represents the generality of supposedly true statements ('synapomorphies') that can be made about the terminal taxa' (Nelson, 1979: 8, footnote 1).

Other commentators have applied alternative words and phrases to the 'generation and legitimation' stages of homology. De Pinna (1991: 373) argued that his terminology (primary and secondary homology) 'seems more appropriate than the others employed so far, because it emphasizes that the two stages are interdependent and complementary, and that any homology hypothesis is necessarily tied to both, at least potentially'. Since primary and secondary homology are interdependent and complementary, we have found it necessary to include comments relevant to both aspects. We have concentrated on the *representation* of primary homology statements, part of de Pinna's primary homology, and on some practical *matters of analysis* pertinent to kinds of representation, part of de Pinna's secondary homology.

Primary homology

Before discussing representation of data and particular matters of analysis, we briefly digress to comment on the first step in the process and the source of all primary homology statements, the identification of similarities. De Pinna (1991: 373) stated that 'A primary homology statement is conjectural, based on similarity, and reflects the expectation that there is a correspondence of parts [of organisms] that can be detected by an observed match of similarities'. Observation of similarity between anatomical structures, behaviour, or particular biological macromolecules is the fundamental first step. It might seem sufficient, even trivial, to say that such comparisons are undertaken by specialists, those who have perhaps developed a reasonable understanding of a particular group of organisms' anatomy, ultrastructure, or whatever. However, it is not trivial to pass over this part of systematic endeavour or to speak deprecatorily of it as 'merely descriptive' (see Brady (1994a, 1994b) for a lucid discussion on the 'descriptive approach' and its scientific basis; see also Grande (1994)). There can be no substitute for detailed study and understanding, as without it there are no primary conjectures to analyse, or at the very least, they may be exceedingly weak. Over a century ago, T.H. Huxley (1888: 116) (in a different context) described such work as '...the drudgery of exhaustive anatomical, embryological, and physiological preparation...' – sound and thoughtful initial work leads to worthwhile primary homology statements. Explicit recognition of the value of such work and its undertaking might well address some of the recent criticisms voiced by more 'orthodox' Hennigian cladists relative to 'computer' cladists (e.g. Wägele, 1994). Even so and paradoxically, 'hypotheses of homology are conjectures whose source is immaterial to their status' (Patterson, 1982: 58; de Pinna, 1991: 373; Nelson, 1994: 123), because it is only through analysis (secondary homology) that they become legitimized. No matter how well data are collected it is unlikely that they will be perfect, but this should not be of an overriding concern as 'recent progress in systematics stems not from better data but from better methods (secondary homology) and more data (primary homology) of roughly the same quality' (Nelson, 1994: 111).

Representation

Conventional coding

De Pinna (1996) characterized primary homology statements as conjectural and went on to note that '[they] may be seen as the way attributes are represented in a matrix of taxa X characters'. The 'taxa-X-character' matrix is a post-Hennig development in cladistics, related to, but not identical with, Hennig's 'argumentation plan' (Hennig, 1966: 91). The representation of primary homology statements in a 'taxa-X-character' matrix has an important implication – the terms 'character' and 'primary homology statement' become one and the same thing conceptually.

Until a few years ago we would have expected little or no disagreement on how to construct a 'taxa-X-character' matrix. In practice, systematic data (similarities) are routinely rendered into series of binary or multistate variables. Consider a simple example using presence/absence data among four taxa, A–D (which might be animals). If taxa C and D share a particular similarity (which might be a backbone) they are each usually given the code '1' and taxa A and B, lacking that similarity (they do not have a backbone), are each given the code '0'. The 0s are merely formalisms representing the complement relation between presence and absence (Patterson, 1982: 48). This relationship is purely logical and devoid of homology content. The 0s have no evidential value – not having a backbone is common to all non-vertebrate life (molluscs, mites, diatoms, bacteria, etc.), but does not indicate what non-vertebrate life is; the only relevant information is that designated by the 1s. If primary homology statements are initial conjectures of homology and are rendered for analysis as a matrix of 'taxa-X-characters' then it is not only empirical observations of similarity that are recorded in such matrices. In addition, at least potentially, is the supposed evidential content of observed similarity; the aspect of primary homology that is useful in determining relationships. This is what is represented by 1s in a 'taxa-X-character' matrix and might best be termed *the cladistic content of primary homology.*

Whatever character coding protocol is adopted (1s and 0s are not compulsory, so some other label might designate shared similarity), each column, or variable, in the 'taxa-X-character' matrix incorporates at least three elements of information: shared similarity (the element that is 'observed' – the empirical part); the notion that everything in the column is comparable (an element of underlying 'sameness'); and a third element, the above-mentioned complement relation (that element we do not observe – the logical part). With respect to conventional binary representation, Brower and Schawaroch (1996) recently segregated de Pinna's primary homology into two parts, 'topographic identity' and 'character state identity'. Topographic identity is nothing more than a notion of fundamental comparability (complete symplesiomorphy in cladistic parlance and what we regard as the element of underlying 'sameness') and is designated by 0s in conventional binary coding. It may appear confusing that symplesiomorphy and the complement relation are both represented by 0s in conventional coding, but this probably is without significance as neither has any homology content at a relevant level. 'Character state identity' is, on the other hand, directly equivalent to a conjectural primary homology statement and is designated with 1s.

Similarity in presence/absence data is directly equivalent to a primary homology statement and to its cladistic content. Recognizing the cladistic content, if any, in other classes of characters is not so straightforward, sometimes requiring analysis itself. For example, macromolecule sequence data for the most part lack the presence/absence class of data (except for insertions or deletions of nucleotide sites, of which see Lloyd and Calder, 1991). Recognizing the shared significant similarity among four taxa, two with adenine (A) at a particular nucleotide site and two with cytosine (C) at the same site, depends entirely on whether the outgroup among the four has A or C, which itself depends entirely on the process of alignment. Whatever the outgroup has, the remaining pair are treated as sharing derived similarity and would be given 1s (or equivalents) in conventional binary coding (but see additional comments below under 'Non-additive multistate characters').

Multistate characters need not be considered different from binary characters as for purposes of computer implementation multistate variables are usually (but perhaps not always) conceived of as series of binary variables, such that all conventional data are ultimately represented and interpreted in binary form (see below).

Three-item representation

Primary homology statements are conjectures of homology, based on similarity, and it is the evidential content (the cladistic content) of this similarity that we seek to represent in a matrix for analysis. Nelson and Platnick (1991) introduced an alternative representation of primary homology statements for the purpose of analysis – the three-item statement. Three-item statement representation presents data not as binary or multistate variables but in terms of the simplest expression of relationship, the three-item statement. For instance, in the example above, if C and D share some feature to the exclusion of A and B, then the three-item statements A(CD) and B(CD) jointly imply that C and D share a relationship to the exclusion of A and to the exclusion of B. In a similar fashion to binary variables, suites of three-item statements can also be arranged in a matrix for analysis, as a 'taxa-X-three-item statement' matrix. Three-item statements represent *directly* the relationships implied by the data, leaving out symplesiomorphous similarity. They are not a 'transformation' of conventional binary representation, as they have been sometimes portrayed (Kluge, 1993; Farris, *et al.* 1995) in spite of the fact that there is a relationship between standard binary and multistate variables and suites of three-item statements (see below). They are direct statements of relationship implied by any shared similarity, and it is conceivable that data could be collected in this way at the lab bench. One needs only a moment's reflection on how data are actually acquired to realize this fact (consider, as one example, the salmon, the lungfish and the cow; Gardiner *et al.*, 1979).

A note on terminology

Nelson and Platnick (1981) presented a detailed account of what they referred to as 'component analysis'. Component analysis is applicable to both geographic (areas) and taxon (characters) analysis. In actuality two different kinds of component analysis were described, both sharing as fundamental the study and analysis of branching diagrams (Page, 1990). Problems relating to branching diagrams can be reduced to their simplest form, the *three-taxon problem*. Three-item analysis (analysis of three-item statements), while having some connection with component analysis (Platnick *et al.* 1996) and the three-taxon problem, is not the same thing as component analysis and the synonymy of the terms 'three-taxon problem' and 'three-taxon statement' as used by some authors (e.g. Page, 1992: 87; Hennipman and Roos, 1983: 327; Wilkinson, 1994a: 344; Lyons-Weiler *et al.*, 1996: 750) is unfortunate (for some further details see Williams, 1996a). It is worth noting that the term 'three-taxon statement' has also been used to represent the basic 'three-taxon problem' (see, for instance, Eldredge and Cracraft, 1980: 50) and is still occasionally used in that fashion today. Nevertheless, component analysis, three-taxon problems, and three-item analysis all deal with the study of branching diagrams, i.e. cladograms.

Three-item analysis has been used for geographic data (Nelson and Ladiges, 1992a; Morrone and Carpenter, 1994) and consensus trees (Nelson and Ladiges, 1994a) as well as for character data (Udovicic *et al.*, 1995; Patterson and Johnson, 1995; Williams, 1996b). The term *three-area statement* is used for geographic data,

Table 9.1 A taxon X three-item matrix for six taxa (ABCDEF), two of which (EF) possess the informative state

Taxon	A(EF)	B(EF)	C(EF)	D(EF)
A	0	?	?	?
B	?	0	?	?
C	?	?	0	?
D	?	?	?	0
E	1	1	1	1
F	1	1	1	1

three-item consensus is used for consensus data and *three-taxon statement* for systematic data, each reflecting the 'units of relationship' or focus of analysis under study. A general term covering all kinds of data is *three-item statement* and a general term covering all kinds of analysis is *three-item analysis*. In the following account, we use the general terms 'three-item statement' and 'three-item analysis'. The usual form of parsimony analysis is referred to as *conventional analysis* with conventional characters and matrices referred to as *standard*.

The taxon X three-item matrix

A single three-item statement includes only three terminals such that a particular three-item statement pertinent to a 12-taxon problem, for example, ignores the nine other terminals (they are included by other statements). Current 'parsimony' computer programs require all cells of a matrix to be filled. This makes the representation of primary homology statements by suites of three-item statements less than straightforward. The requirement that all cells in a matrix must be filled may be satisfied by the use of a question mark (?) in those not immediately applicable to any particular statement. Table 9.1 provides an illustration for the relation ABCD(EF).

The ? value in cells of a taxon-X-three-item matrix should be interpreted as having the meaning 'inapplicable', as opposed to 'either/or' or 'polymorphic'. This simple heuristic device has been persistently misunderstood by critics (Harvey, 1992; Kluge, 1993, 1994) who have interpreted it as somehow creating missing data values (see 'Minimal tree and node support' below). It does no such thing (Platnick, 1993). As Nelson (1992, 1993, 1996) has repeatedly pointed out, for the relation ABCD(EF) the data are A(EF) + B(EF) + C(EF) + D(EF) = ABCD(EF). The '? problem' would be eliminated if, in future, programmers allowed an option for matrix creation to equal 'no data'. As far as we can tell this is still not available in the NEXUS file format (Maddison *et al.*, 1997: 599).

Whenever observed similarity is present among more than just two taxa, e.g. A(BCD) or a multistate character, A(B(CDE)), there is logical dependency among the implied three-item statements that must be corrected. For example, the binary character expressing the relationship A(BCD) has three statements, A(BC), A(BD) and A(CD). Of the three statements, any two combined logically imply the third: A(BC) + A(BD) = A(BCD); A(BC) + A(CD) = A(BCD); and A(BD) + A(CD) = A(BCD). This redundancy is addressed by 'weighting' each statement by its absolute value, taken

Table 9.2 Hennig86 matrix for character A(B(CDE)

```
Xread
9          6
OUT   0    0    0    0    0    0    0    0    0
A     0    0    0    0    0    0    ?    ?    ?
B     1    1    1    ?    ?    ?    0    0    0
C     1    ?    ?    1    1    ?    1    1    ?
D     ?    1    ?    1    ?    1    1    ?    1
E     ?    ?    1    ?    1    1    ?    1    1
;
CCode
      /3 0  /3 1  /3 2  /2 3  /2 4  /2 5  /2 6  /2 7  /2 8
proc/;
```

as the ratio of the number of independent statements, $(n-1)(t-1)$, to the total number of implied statements, $(t-n)n(n-1)/2$. For binary characters this ratio equals $2/n$; thus for A(BCD) each statement has an absolute value of 2/3. This 'correction' is known as *fractional weighting* and can be implemented using current 'parsimony' computer programs by assigning appropriate weights (Nelson and Ladiges, 1992b). Redundancy is addressed in this way rather than by attempting to omit any particular statement (as seems to be suggested by Deleporte (1996: 281) and De Laet (1997: 47–58)) as all implied statements remain a property of the data. The effect of fractional weighting for the relationship A(BCD) is to reduce the summed weight of the three statements. The summed weight of the fractionally weighted statements is 2; the summed weight of the three uncorrected statements is 3.

Implementing fractional weights to correct for dependency among statements is not always straightforward if using Hennig86 for subsequent analysis. For example, the Hennig86 matrix for the character A(B(CDE) is given in Table 9.2. There is an absolute value weight of 1.000 for the first three statements and 0.667 for the subsequent six, correcting for the dependency inherent in the (CDE) node of the character. A rounding factor must be selected that achieves the relative values of the combined fractional weighting scheme and does not exceed the maximum limit imposed for the total weight of a matrix. A rounding factor of 3 was applied (the common denominator of 1 and 2/3) to produce whole-number weights that achieve the relative values of the fractional weights of 1.000 and 0.667, for a total matrix weight of 21 (the sum of the weights).

The MS-DOS program TAX (Nelson and Ladiges, 1994b) renders a matrix of standard characters as a matrix of three-item statements, creating appropriate output for analysis with PAUP, Hennig86 or NONA. The TAX suite of programs consists of the following: MATRIX (converts Hennig86 formatted files into an appropriate format for use with TAX), MOMATRIX (converts molecular data sets in Hennig86 formatted files into an appropriate format for use with TAX), TAX (converts data into three-item matrices; options exist for using files generated from MATRIX or entering data directly; output can be in either Hennig86 or PAUP file format), CONPAUP (converts PAUP treefiles into Hennig86 treefiles) and TAXUTIL (utilities for Hennig86 treefiles: numbers of statements, numbers of nodes, creates data files for use with three-item consensus). There are also programs for dealing with areas in biogeography (TAS, Nelson and Ladiges, 1994c). Complicated characters and large numbers of taxa may produce many thousands of implied three-item statements, resulting in a matrix too large for analysis with Hennig86. Use of NONA or PAUP for analysis may be the only practical recourse for these situations at present.

Why three?

Representation of primary homology in a matrix of three-item statements has been questioned by suggesting that some other number of statements, e.g. 4-, 5-, or n-item statements, might be as useful as, or perhaps even more useful than, three-item statements (De Laet (1997) and discussion in Wilkinson (1994a) from a consensus tree viewpoint). This suggestion does not address the fundamental issue of just what aspect of primary homology should be rendered as a matrix entry for analysis, because entertaining the idea of n-numbered statements accepts the basic premise of representing primary homology statements as relationship. Moreover, the unit value of relative relationship is a statement of relationship among three items, as smaller statements (two- or one-item statements) convey no information on relative relationship (but see Kluge, 1994: 411). Furthermore, larger statements (n-item statements) can be represented by a suite of three-item statements. For example, A(BCD) can be represented as A(BC) + A(BD) + A(CD), or by any pair of statements among the three possible from A(BCD). Three-item statements are the most general form of representation.

Binary characters

The relationship between standard binary characters (a conventional character with two states) and three-item statements is illustrated with the following formula:

$$(t - n)n(n - 1)/2$$

where t = the total number of taxa and n = the number of taxa with the informative (apomorphic) state (Nelson and Ladiges, 1992b). For instance, data expressing the relationship ABC(DE) have three statements: A(DE), B(DE) and C(DE) ($n = 2$, $t = 5$, hence $(5 - 2)2(2 - 1)/2 = 3$); data expressing the relationship AB(CDE) have six statements: AB(CDE) = A(CD), A(CE), A(DE), B(CD), B(CE), B(DE) ($n = 3$, $t = 5$, hence $(5-3)3(3-1)/2 = 6$). This formula can be used to calculate the number of three-item statements implied by any 'standard' binary variable.

When $n > 2$ there is logical dependence among statements (Nelson and Ladiges, 1992b) that must be corrected by applying fractional weighting. Redundancy resulting from state distributions within a conventional multistate character is dealt with in the same way (see below).

Additive multistate characters

Multistate characters represent data with more than one node and, for additive characters, are usually understood as suites of binary characters for the purpose of

Table 9.3 Comparison of uniform and fractional weighting with respect to one multistate character and its binary equivalents (after Nelson and Ladiges, 1992b and Kitching *et al.*, 1998)

	Uniform weighting		Fractional weighting		Multistate
	Binary Ch 1a A(BCD)	Binary Ch 1b AB(CD)	Binary Ch 1a A(BCD)	Binary Ch 1b AB(CD)	Ch 2 A(B(CD)
A(BC)	1	–	2/3	–	1
A(BD)	1	–	2/3	–	1
A(CD)	1	1	2/3	1	1
B(CD)	–	1	–	1	1
	3 + 2		2 + 2		4
Total	5		4		4

analysis. It has been recognized for some time that such re-coding may involve redundancy, an idea recently revived in the exchange between Purvis (1995) and Ronquist (1996). Kluge (1993: 248) has taken the opposite view and suggests that '. . .the additive binary coding method (Farris *et al.*, 1970: 180), which transforms additive multistate characters into sets of binary characters, is an exact operation. . .'.

From the perspective of three-item analysis, '[A] multistate character equals a suite of unique three-taxon statements (no statement appears more than once in the suite)' (Nelson and Ladiges, 1994b). For instance, the multistate character expressing the relationship A(B(CD)) has four statements: A(BC), A(BD), A(CD), and B(CD). When converted into binary equivalents, A(B(CD)) yields two characters: A(BCD) and AB(CD) (Table 9.3). A(BCD) yields three statements, A(BC), A(BD), and A(CD), and AB(CD) yields two statements, A(CD) and B(CD), giving a total of five statements. The two binary characters have one more statement than the single multistate character, as A(CD) occurs twice and is considered redundant with respect to the original multistate character (Table 9.3). Fractional weighting reduces the total values giving a total weight of 4, as in the multistate character. However, the proportional weights differ and both uniform and fractional weighting of suites of binary characters differ from the equivalent multistate character.

Of further interest is a coding procedure suggested by Purvis (1995) and explored by Ronquist (1996: 248, Fig. 2). Purvis suggested his coding as a way of representing cladograms rather than characters, but the principle remains the same. Purvis noted that there was redundancy in additive representation of cladograms with more than one node (= multistate characters). He suggested a protocol to remove the redundancy by coding taxa at each node with a 1, scoring taxa sister to that node with a 0 and scoring all remaining taxa with a question mark.

The revised coding Purvis (1995) suggested was considered earlier by Nelson and Ladiges (1992b). For instance, the ordered multistate character (A(B(CDE) can be represented by two binary characters, A(BCDE) and AB(CDE). Each binary character yields six statements: character A(BCDE) yields A(BC), A(BD), A(BE), A(CD), A(CE) and A(DE) and character AB(CDE) yields B(CD), B(CE), B(DE), A(CD),

Table 9.4 Comparison of numbers of three-item statements derived from the cladogram (= multi-state character) (A(B(C(DE)). Sts = three-item statement; standard coding = component coding = binary coding; Purvis (1995) coding as described in the text; weight refers to either uniform weighting (UW, all statements have equal weight) or fractional weighting (FW, some statements have fractional weights)

Sts	Standard coding weight		Purvis coding weight		Multistate weight	
	UW	FW	UW	FW	UW	FW
A(BC)	1.000	0.500	1.000	0.500	1.000	
A(BD)	1.000	0.500	1.000	0.500	1.000	
A(BE)	1.000	0.500	1.000	0.500	1.000	
A(CD)	2.000	1.167	1.000	0.500	1.000	
A(CE)	2.000	1.167	1.000	0.500	1.000	
A(DE)	3.000	2.167	1.000	0.500	1.000	
B(CD)	1.000	0.667	1.000	0.667	1.000	
B(CE)	1.000	0.667	1.000	0.667	1.000	
B(DE)	2.000	1.667	1.000	0.667	1.000	
C(DE)	1.000	1.000	1.000	1.000	1.000	
Totals	15.000	~10.000	10.000	~6.000	10.000	

A(CE) and A(DE), giving a total of 12 statements. Three statements appear twice, A(CD), A(CE) and A(DE). Using the coding suggested by Purvis, there are two modified binary characters, A(BCDE) and B(CDE), yielding a total of nine state-ments: A(BCDE) yields six statements, A(BC), A(BD), A(BE), A(CD), A(CE) and A(DE), and B(CDE) yields three statements, B(CD), B(CE) and B(DE). Both ways of coding will give the correct result when analysed using a parsimony program but, as Nelson and Ladiges (1992b) point out, the representation of data using Purvis' (1995) coding 'is inaccurate for what it implies about the information content of each state...state 1 [A(BCDE)] seems twice as informative as state 2 [AB(CDE)], when the reality is just the opposite'.

Examination of the multistate character (A(B(C(DE)) provides a further aspect. The cladogram (A(B(C(DE)) can be represented by three binary characters (= three components), A(BCDE), AB(CDE) and ABC(DE), which yield a total of 15 three-item statements (Table 9.4). Among those 15 statements, 10 are distinct, with four of those 10 occurring more than once. Application of fractional weighting effec-tively reduces the relative value of the 15 statements, yielding a total weight of 10, reflecting the total weight of the distinct statements (Table 9.4). Analysis of the cladogram (A(B(C(DE)) as a multistate character yields 10 unique statements, with both uniform and fractional weighting (Table 9.4). If the same cladogram (A(B(C(DE)) is coded according to Purvis' method, the three binary characters become modified to A(BCDE), B(CDE) and C(DE), yielding a total of 10 state-ments (Table 9.4). However, while the 10 statements derived from these binary characters are identical to those from the multistate character, they are not all logi-cally independent, unlike those from the multistate character, thus they require correction using fractional weighting. Application of fractional weighting yields an overall total weight of ~6 (Table 9.4). This reduced total weight is due in part to the missing values added for each binary character and the consequent loss of informative statements using separate components. One might conclude that Purvis' method of coding is more accurate as it corrects for redundancy without further weighting. However, it remains inaccurate as the information represented is significantly reduced in comparison either to the three separate binary characters or to the single multistate character. In effect Purvis' coding implies that it is inappropriate to compare the advanced states of a multistate character to basal taxa; it acts as equivalent to the direct three-item representation of a multistate character but suffers from dependency among resultant statements.

The issue of the inappropriateness of comparing advanced states of a multistate character to those in basal taxa might be made clearer by thinking about a less abstract example. (We thank Norman Platnick for suggesting the example.) Consider that among taxa ABCDE, taxon A has no stalk, taxa BCDE have developed a round stalk, in taxa CDE the stalk has been modified to be triangular in cross-section, and that taxa DE have developed bulbous expansions at the corners of the triangle. Purvis' coding avoids counting A(DE) more than once, as happens with standard binary representation of this multistate character (Table 9.4). The method accurately represents A(DE) because A can hardly have a triangular stalk, or one with bulbous expansions, if it does not have a stalk at all. Purvis' coding treats the modification of the stalk into one that is triangular in cross-section (CDE) and the development of bulbous expansions (DE) as inapplicable to A. With regard to comparing Purvis' coding with direct three-item representation of multistate characters, it might be possible to pursue the differences in actual kinds of characters with respect to dependency among terminals and to decide whether, for instance, particular sets of observations require consideration of dependency.

In general, there is a real difference (in terms of information) among one multistate character, its equivalent binary characters and its equivalent corrected binary characters, in spite of the fact that they yield the same result; equivalence of cladogram and matrix may not be quite as straightforward as some would like. Three-item representation provides a means to distinguish between suites of (congruent) binary characters and single additive multistate characters in terms of their overall information content. Three-item statement representation avoids the redundancy inherent in conventional or modified additive binary coding of multistate characters (Tables 9.3, 9.4) and identifies non-independence in the data, whereas conventional analysis does not.

Non-additive multistate characters

Nelson and Ladiges (1992b) did not address the issue of non-additive (unordered) multistate characters. Commonly used in molecular studies, non-additive characters are optimized using Fitch's 'minimum mutation' model of character state change (Fitch, 1971; Mickevich and Weller, 1990: 168). Conventionally, non-additive characters have been used to avoid imposing any kind of order on the states of multistate characters. In effect, no relations among taxa are specified. According to some commentators, Fitch optimization 'is not generally satisfactory' (Platnick, 1989b: 22) or even 'nihilistic' ('Fitch optimization. . . is the most parsimonious, most conservative (even nihilistic) procedure' (Wheeler, 1990: 269)).

Table 9.5 One multistate character and its three-item equivalent (after Hawkins *et al.*, 1997: 277; their Table 3, and Table 2, analysis 2): A–E are taxa, AO is equivalent to the outgroup used by Hawkins *et al.* (1997), the assumption being that AO is not part of the problem to be solved

Taxa	Multistate	Three-item	
AO	0	0	0
A	0	0	0
B	1	1	?
C	1	1	?
D	2	?	1
E	2	?	1

Three-item analysis sees 'non-additive' characters somewhat differently. The values specified by the variable states are the *only* indicators of relationship and the minimal tree for these relations usually lacks a basal node. For instance, the multistate character in Table 9.5 contains information relevant to only B+C and D+E.

A recent example discussed by Hawkins *et al.* (1997: 277; their Table 3, and Table 2, Analysis 2; our Table 9.5) is of interest in this context. Parsimony analysis using conventional non-additive multistate coding yields four cladograms: A(BC(DE)), A(DE(BC)), A((BC)(DE)), and A(BC)(DE).

For various reasons, Hawkins *et al.* (1997) reject the solution A(BC)(DE). This is the solution found by three-item analysis and requires further consideration. Analysis of the three-item data in their Table 3 yields 11 trees, all of which are not minimal. Solution A(BC)(DE) is the minimal tree for these data; for the rationale see our Table 9.6, example 6 (after Nelson and Ladiges, 1992a: Table 6, example 6) and Nelson and Ladiges (1996: 44). In terms of three-item analysis, the minimum information required for the solution A(BC)(DE) is two statements: A(BC) + A(DE). The total number of statements implied (or possible) from this solution is six: A(BC) and A(DE) (those required), and B(DE), C(DE), D(BC) and E(BC) (those implied).

Hawkins *et al.* (1997: 280) favour the solution A(BC(DE)) or A(DE(BC)) on the basis that it accords better with Platnick's (1979) notion of character states as additions rather than alternatives. Consider the following possibilities (see also Nelson and Ladiges, 1992a: table 6, example 6 and Nelson and Ladiges, 1996: 44). If the minimal tree for A(BC) + A(DE) is A(BC)(DE) then addition of single, but different and non-conflicting, statements to this solution obtains various results. The addition of either statement A(BD), A(BE), A(CD) or A(CE) to A(BC) + A(DE) yields the solution A(BCDE). Thus it is information from outside the set (relations with A) that resolves the basal node relative to the separate nodes of A(BC) or A(DE). Only addition of statements from *inside the set* BCDE will resolve the solution(s) similarly to Hawkins *et al.* (1997):

A(BC) + A(DE) + **B(DE)** = A(C(B(DE)))

A(BC) + A(DE) + **C(DE)** = A(B(C(DE)))

or

A(BC) + A(DE) + **D(BC)** = A(E(D(BC)))

A(BC) + A(DE) + **E(BC)** = A(D(E(BC)))

and addition of four statements, with conflict, yields their favoured solution:

A(BC) + A(DE) + **B(DE)** + **C(DE)** = A(BC(DE)).

Hawkins *et al.* (1997) reject the solution A((BC)(DE)) on grounds that it is over-resolved for the data at its basal node; the additional basal node for (BC)(DE) is spurious. This indeed seems to be the case. NONA finds only three trees, A(BC(DE)), A(DE(BC)) and A(BC)(DE). From the perspective of three-item analysis two extra statements are required for that solution:

A(BC) + A(DE) + **B(DE)** + **D(BC)** = A((DE)(BC))

A(BC) + A(DE) + **C(DE)** + **E(BC)** = A((DE)(BC))

This indicates that each subset is related directly outside the set as well as to the other set, which is a greater degree of precision.

Relating this to Hawkins *et al.*'s (1997) 'real' observations, BC is 'red tail' (or 'red') and DE is 'blue tail' (or 'blue'). (We avoid the very real issue here concerning what exactly is homologized with respect to colour. Is 'red' truly a homologue of 'blue'?) What data are present to relate DE to BC? Only 'tail' which all the organisms B–E are said to have, and which is therefore irrelevant (symplesiomorphy; = AB(CDEF); lacks cladistic content at an appropriate level). Also irrelevant is that DE and BC relate to organisms without tails, A(BD), A(BE), A(CD) and A(CE), as each statement confirms they have tails and renders colour irrelevant (complement relation; lacks any cladistic content).

While Hawkins *et al.* (1997) did not wish to support this particular multistate character representation from among their examples, it is instructive relative to the standard treatment of molecular characters. Suppose that BC is represented by guanine (G) and DE by thiamine (T). What data are present to relate DE (G) to BC (T)? Only 'column x or A' (symplesiomorphy). It is irrelevant that DE (G) and BC (T) relate to organisms without G or T (statements A(BD), A(BE), A(CD) and A(CE)) as each statement confirms they have T (present in the column) and renders either G or T irrelevant. This example is examined not with the desire to criticize the general conclusions of Hawkins *et al.*, but simply to point to the inherent ambiguity of standard binary and multistate 'formalisms'; in effect each coding procedure Hawkins *et al.* suggest relates the taxa in one and the same way, AB(CD)(EF).

Secondary homology

Three-item analysis

Problems that interest systematists are usually of a reasonable size, yet even with a relatively small amount of data problems can result when a taxa-X-three-item matrix has many thousands of entries. With appropriate cautionary measures, currently available parsimony programs can be used to summarize these matrices. To be sure, using current parsimony programs for three-item analysis is something of a 'misuse', but it is nevertheless appropriate because the optimal tree for a suite of three-item statements can, in principle, be achieved because, if nothing else, tree length is exact (Nelson, 1992, 1993). In other words, the shortest tree for a three-item matrix is the optimal tree (but see below).

Much has been said about the 'optimization' of three-item statements. All available parsimony programs optimize characters on a cladogram, but the fact that current computer programs are required to optimize characters is irrelevant when undertaking a three-item analysis. We repeat (see also Kitching et al., 1998; Nelson, 1992, 1993; Platnick, 1993; Siebert and Williams, 1998), optimization is irrelevant to three-item analysis. Some of these ideas are expanded upon below.

Tree length and three-item statements

The relationship between tree length and three-item statements is exact and can be expressed as follows (Nelson, 1992; Nelson and Ladiges, 1994a):

$$\text{tree length (L)} = \Sigma \text{ ATS (accommodated three-item statements)}$$
$$+ \Sigma \text{ } 2 \times \text{NTS (not accommodated three-item statements)}$$

'Accommodated' simply means that a statement 'fits to one node exactly' (Platnick, 1993) and is not a new 'measure' (Kluge, 1994: 411). As in conventional analysis, the shortest trees are the 'best' trees; shortest after redundancy has been taken into consideration.

Minimal trees and node support

Analysis of a three-item matrix using current parsimony programs may yield one or more cladograms. If a three-item matrix yields more than one cladogram, all may not be minimal in the sense that all resolved nodes may not be supported by data (similar problems occur with matrices of standard characters; see Coddington and Scharff (1994), Nixon (1996), Nixon and Carpenter (1996), and Wilkinson (1995)). In this sense, a minimal tree is the shortest tree with all resolved nodes supported by data. For instance, with 15 taxa of which only six have the informative state, there are 135 statements. Analysis of these statements using Hennig86 yields 3635+ possible solutions (the number of trees is not exact but represents the limits of the computer used for this analysis). Of the 3635+ solutions, none have all nodes fully supported by data but all have length of 135 (the number of nodes in the cladograms varies from five to 13). The tree length equals the total number of three-item statements as all are accommodated on every one of the trees. In this case, the strict consensus tree of the 3635+ cladograms is also a shortest cladogram but with one resolved node that is fully supported by the data (uniting at one node all six taxa with the informative state); this is the preferred cladogram, as it is minimal for nodes and data. This observation applies to all three-item representations of single node relationships ('binary' characters) where $n > 2$ (Nelson

Table 9.6 The first column gives the example number after Nelson and Ladiges, 1992a: Table 6; the second column gives the data as pairs of three-item statements, A(BC) + one other, the third, fourth, fifth and sixth columns are the numbers of trees recovered from analysis using Hennig86, NONA, with (+) and without (−) ambiguous optimizations and compatibility, respectively. The seventh column is the 'correct' solution (equivalent to 'assumption 2' solutions for biogeography). NONA produces one extra tree (a bush and hence not minimal) for examples 5 and 6, three extra non-minimal trees for example 7, and seven extra non-minimal trees for example 8. The protocol for determining minimal trees followed that outlined by Nixon and Carpenter (1996: 311).

Example		Programs				Minimal tree A2 solution
		Hennig86	NONA (+)	NONA (−)	Compat.	
	Four taxa					
1	A(BC) + D(BC)	3	1	1	3	AD(BC)
2	A(BC) + A(CD)	3	1	1	3	A(BCD)
3	A(BC) + C(BD)	1	1	1	1	A(B(CD)
4	A(BC) + B(AD)	1	1	1	1	(AD)(BC)
	Five taxa					
5	A(BC) + D(CE)	11	11	4(3)	7	AD(BCE)
6	A(BC) + A(DE)	11	9	3(2)	14	A(BC)(DE)
7	A(BC) + B(DE)	7	7	7(4)	7	A(BC)(DE)
	Six taxa					
8	A(BC) + D(EF)	55	47	18(11)	27	AD(BC)(EF)

and Ladiges, 1993). In general, the minimal tree may be found in many cases by the strict consensus of all most parsimonious cladograms; when the strict consensus tree is longer than the most parsimonious cladograms, it is not minimal (Nelson and Ladiges, 1993: Table 1).

To achieve a more accurate result initially, one may use the parsimony program NONA (Goloboff, 1993). NONA allows ambiguous optimizations to be suppressed and results so far indicate that it usually, but not always, finds the minimal tree in a three-item matrix more efficiently than Hennig86. For three-item matrices equivalent to single binary characters, the minimal tree is always found. A more significant series of examples is illustrated by re-analysis of those presented by Nelson and Ladiges (1992a).

Inspection of Table 9.6 illustrates results when pairs of statements are combined where one statement is A(BC). The correct answer is easy to discover by simple inspection and addition of the statements (the correct solution corresponds to the 'Assumption 2' (A2) solution relevant to biogeographic analyses (Nelson and Ladiges, 1992a; see also Platnick *et al.*, 1996)). Inspection of column 3 in Table 9.6 (Hennig86) shows that when the data are coded for analysis using question marks for no data (*not* missing data) there is an over-production of trees in six of the eight examples. For instance, A(BC) + D(BC) (Table 9.6, example 1) yields three cladograms, two of which are irrelevant. In each of the examples with four taxa (Table 9.6, examples 1–4), NONA, with the unambiguous optimization default active (command *amb-*), produces the correct answer, eliminating those cladograms with over-resolved nodes. The situation differs with cases involving five and six taxa. Although NONA produces fewer cladograms than Hennig86 in most cases, it still does not always find the single correct minimal solution. At present no parsimony program will always find the minimal tree in these simple examples. It is worth noting that the 'correct' solution is often found among the many solutions recovered by both Hennig86 and NONA (Nelson and Ladiges, 1992a).

Example 6 in Table 9.6 presents a further parameter in obtaining the correct minimal cladogram, relevant to the non-additive multistate character example above. For the two statements A(BC) + A(DE), Hennig86 produces 11 trees, NONA (with ambiguous optimization active) produces nine trees and NONA (with unambiguous optimization active) produces two cladograms: A(BC)(DE) and A(BCDE). (In actual fact, NONA produces three cladograms. The third cladogram, not discussed above, is a collapsed bush.) The latter two cladograms from NONA differ in that one has two nodes, the other has one node. Intuitively, one might suspect the one-node cladogram to be minimal, or at least as good as the alternative A(BC)(DE). However, when the accumulation of more data relevant to this solution is considered, the situation is somewhat different (see Nelson and Ladiges (1996: 22) and the multistate character example in Table 9.4). For the cladogram A(BC)(DE), addition of further statements that do not conflict with A(BC) and A(DE) results in the cladogram A(BCDE). For instance, if A(BE) is added to A(BC) + A(DE), the solution is A(BCDE). Conversely, there are no data that could be added to the cladogram A(BCDE) to arrive the solution A(BC)(DE). Thus the one-node cladogram, A(BCDE), requires *more* data than the two-node cladogram and is, in this sense, minimal. In this respect, cladograms resulting from three-item analysis need closer inspection than simply accepting the suite of most parsimonious trees or indeed simply inspecting the strict consensus of all cladograms produced. Such situations help to focus on exactly what data do support nodes in the resulting trees. An appreciation of how three-item analysis deals with increasing resolution with increasing data points again to its greater precision.

Wilkinson (1994b) suggested that 'parsimony' analysis of a three-item matrix should yield identical results to a compatibility analysis. In other words, the 'correct' solution can be arrived at by the use of either a compatibility or a parsimony program. Wilkinson (1994b: 222) provided three examples of three-item analysis that violates the 'pairwise compatibility theorem' (Fitch, 1975). His examples are combinations of pairs of three-item statements drawn from a selection of three: D(AB), B(AC) and C(AD). Each permutation yields one tree: D(AB) + B(AC) = D(B(AC), D(AB) + C(AD) = C(D(AB), and B(AC) + C(AD) = B(C(AD). Processed with a parsimony program (Hennig86), results are the same (a single tree for each combination). Wilkinson concluded that because optimization of the missing value in each permutation requires a different value each time, the statements are incompatible, and because '...missing entries invalidate the pairwise compatibility theorem...it limits its applicability to three-item statements of the character-based approach to clique analysis' (Wilkinson, 1994b: 223).

His logic is the same as that applied to the implementation of parsimony analysis: because a program does something, the non-computed conclusion must be incorrect. For the sake of clarity, we repeat: the question mark in the fourth item for one three-item statement in a four-taxon matrix means nothing; there is no point in attempting to optimize the value; it matters not if the program suggests that they are contradictory (0 for C in one tree, 1 for C in another tree); the program is working with an entry that is nothing more than a formalism. Of interest, though, is that compatibility programs find the same answer as parsimony programs in Wilkinson's examples. However, it is of even greater interest to compare the results from compatibility analyses with those from parsimony analyses using the examples in Table 9.6: examples 5, 6 and 8 differ in the number of trees produced. Examples 2 and 5 do not contain the correct ('Assumption 2') solution among the results. This seems to suggest that compatibility programs do not necessarily find minimal trees in three-item analyses.

Optimization

Having established the relationship between the optimal three-item tree and arriving at that solution using current parsimony programs, we can return to the issue of optimization, briefly touched on above. Current parsimony programs, because of their design, may display ambiguous optimizations for some three-item statements. The programs treat each statement (column in the matrix) as a 'conventional' character and are required to assign values to each taxon on every node in spite of having only three 'real' values. Consider the cladogram A(B(C(DE)) and the three-item statement A(CD). When A(CD) is 'optimized' on to A(B(C(DE)), parsimony programs will assign '0/1' to node BCD and '1' to node CDE and DE. Note that assignment of '0/1' to node BCD is irrelevant as the statement fits node CDE exactly: the relationship A(CD) is a true statement of relationship relative to the cladogram A(B(C(DE)). The only reason programs suggest ambiguity is that they are designed to deal with the conventional situation – even with conventional analysis things are not quite as straightforward (see, for instance, the discussion in Nixon (1996: 368–370) with reference to conventional analysis).

Kluge (1994: 407) suggested that there might be 'multiplication of information' when data are represented by suites of three-item statements. He considered the relationship between a conventional matrix with three characters and their three-item equivalents. The conventional matrix contained data for three characters – A(BCDE), AB(CDE) and ABC(DE) – which together yield the solution A(B(C(DE)). The same data, seen as three-items, yield 10 statements (ABC × 1, ABD × 1, ABE × 1, ACD × 2, ACE × 2, ADE × 3, BCD × 1, BCE × 1, BDE × 2 and CDE × 1); all 10 statements exactly fit (one step each) the cladogram A(B(C(DE)). Kluge (1994: 407) directed his attention to three statements which fit certain nodes exactly but have ambiguous optimizations in the conventional sense. The statements he dealt with are A(BC), A(CD) and A(DE). If they are considered on their own, as Kluge did, results differ from when one considers the entire suite of 10 statements. Kluge concluded that A(BC) + A(CD) + A(DE) does not result in the A(B(C(DE))) cladogram of Nelson (1993: 261) but should result in A(BCDE) as implied by the 'original' character. It should be straightforward to see that A(BC) + A(CD) + A(DE) = A(BCDE), (see Nelson and Ladiges, 1992b: Table 6; Nelson, 1993: 264) not A(B(C(DE))).

When one reads a three-item statement on a cladogram it is in the form of 'C and D share a closer relationship to each other than either does to A', not in the form of conventional analysis which sees nodes as possible (ancestral) transformations of one character (or state) into another. A three-item statement is the *relationship* implied by similarity between two of three taxa; the statement either fits a particular tree or it does not. Using an example from Platnick (1993), Farris *et al.* (1995: 217) noted that while one of Platnick's trees fits 65 statements, it 'explains' only 45 in the conventional sense of optimization. The tree in question is (RST)(A(U(B(V(C(W(D(X(E(Y(FZ) and the relationship requiring explanation is RSTUVWXYZ(ABCDEF) which yields 135 statements (listed in Kluge, 1994: 410–11). The considered tree accommodates both R(AF) and X(EF). According to Farris *et al.* (1995: 215), 'Explaining the shared 1s of *a* and *f* by inheritance demands that all 11 stems between *a* and *f* have state 1. But to account for the 0 of *x*, every ancestor of *x* – including eight of those same stems – must have state 0 instead.' In other words, three-item statements do not conform to the usual kind of optimization. Inspection of the tree demonstrates clearly that both R(AF) and X(EF) are true statements of relationship:

R(AF) – (**R** S T)(**A** (U (B (V (C (W (D (X (E (Y (**F** Z)

X(EF) – (R S T)(A (U (B (V (C (W (D (**X** (E (Y (**F** Z)

Similar inspection of all 65 statements illustrates that they all represent true statements of relationship derived from the data; one simply needs to read the statement correctly: '*x* and *y* share a closer relationship to each other than either does to *z*', all statements are then 'explained'; or in the above case 'A and F share a closer relationship to each other than either does to R' and 'E and F share a closer relationship to each other than either does to X'. The attempts by Farris *et al.* to consider ambiguous optimizations of three-item statements in the same fashion as standard characters misses the point (as does Kluge, 1993: 250; Kluge, 1994: 405; Deleporte, 1996). In short, optimization is irrelevant to the result, notwithstanding De Laet's (1997: 58–63) efforts to 'correct' the problem.

Fractional weighting

For fractional weighting, the shortest trees are those that fit the greatest total weight, which need not be the same as the tree that fits the greatest number of statements (Nelson, 1993; Nelson and Ladiges, 1994a). As redundant statements may offer spurious information, one would expect fractionally weighted data to be more precise (Nelson and Ladiges, 1994a and see below).

Three-item examples revisited

Nelson and Platnick (1991)

Nelson and Platnick (1991) presented their initial discussion of three-item analysis using uniform weights and exact tree output from Hennig86 (although they did refer to the possibilities of fractional weighting (Nelson and Platnick, 1991: 363); fractional weighting was developed later, (Nelson and Ladiges, 1992b)). Of the ten examples they presented, the use of fractional weighting renders one example

Table 9.7 Examples from Nelson and Platnick (1991) with results from considering fractional weighting and minimal trees. Binary and three-item matrices follow Nelson and Platnick (1991). Binary matrix 3 as matrix 1 with additional character, matrices 7 and 8 based on data from Carpenter (VESP data set in Platnick, 1987, 1989b). Matrix 7 is the full matrix, matrix 8 is with autapomorphies removed and the matrix on which the three-item result is based. Analysis of matrix 8 yields three trees, each of length 39, when an all-zero outgroup is added. A strict consensus of these three results in a tree also of length 39 but with the terminal trichotomy collapsed. This is the minimal tree for the data; matrix 13 as matrix 11 with additional character; example i is matrix 17 minus character 4 and matrix 18 is minus statements equivalent to character 4

Example	Binary matrix	Three-item matrix	Results		
			Trees	Conventional	Three-item
a	1	2	Fewer	2 cladograms	1 cladogram
b	3	4	Fewer	3 cladograms	1 cladogram
c	5	6	More	2 cladograms	4 cladograms
d	7 and 8	9	Different	1 cladogram	1 cladogram
e	11	12	Same	1 cladogram	1 cladogram
f	13	14	Different	1 cladogram	1 cladogram
g	15	16	Different	1 cladogram	1 cladogram
h	17	18	Different	1 cladogram	2 cladograms
i	17a	18a	Different	1 cladogram	2 cladograms
j	19	20	Different	1 cladogram	1 cladogram

Table 9.8 Standard data for matrix 11 and 13 from Nelson and Platnick (1991)

Matrix 11		Matrix 13	
Taxa	Characters	Taxa	Characters
A	11000	A	110001
B	10111	B	101110
C	01111	C	011111
D	00010	D	000100

(example e in Table 9.7) irrelevant as it yields an identical solution to conventional analysis. Of the nine remaining examples, three include the conventional analysis solution (examples a, b and h), two recover solutions more fully resolved than the conventional analysis (examples c and d), and four recover solutions different to conventional analysis (examples f, g, i, j). Below we examine further those examples that recover different solutions.

Examples e and f

(Matrix 13 of Nelson and Platnick, 1991; = their Matrix 11 + an additional BD(AC) character; Table 9.8.) With uniform weighting, matrix 11 yields two trees; with fractional weighting, matrix 11 results in one tree identical to conventional analysis.

When represented as three-item statements, the data in matrix 11 (example e in Table 9.7, data in Table 9.8) has a total of 11 statements, eight of which are

Table 9.9 Uniform and fractional weighting for matrix 11: UW = uniform weighting, FW = fractional weighting, actual W = actual weight assigned by the Ccode value of Hennig86, × 10 is the rounding factor applied (see text)

| Statement | Matrix 11 (example e) | | | |
	UW	Actual W	FW	Actual W x 10
A(BC)	3.000	3	2.667	27
A(BD)	1.000	1	0.667	7
A(CD)	1.000	1	0.667	7
B(AC)	1.000	1	1.000	10
C(AB)	1.000	1	1.000	10
D(AB)	1.000	1	1.000	10
D(AC)	1.000	1	1.000	10
D(BC)	2.000	2	1.000	20

Table 9.10 Uniform and fractional weighting for matrix 13: UW = uniform weighting, FW = fractional weighting, actual W = actual weight assigned by the Ccode value of Hennig86, × 10 is the rounding factor applied (see text)

| Statement | Matrix 13 (example f) | | | |
	UW	Actual W	FW	Actual W × 10
A(BC)	3.000	3	2.667	27
A(BD)	1.000	1	0.667	7
A(CD)	1.000	1	0.667	7
B(AC)	**2.000**	2	2.000	20
C(AB)	1.000	1	1.000	10
D(AB)	1.000	1	1.000	10
D(AC)	**2.000**	2	2.000	20
D(BC)	**2.000**	2	1.000	20

unique; A(BC) is represented three times and D(BC) is represented twice. Tree (D(A(BC) accommodates the four statements A(BC) × 3, D(AB) × 1, D(AC) × 1 and D(BC) × 2, representing a total of seven statements. Tree (A(D(BC) accommodates the four statements A(BC) × 3, A(BD) × 1, A(CD) × 1 and D(BC) × 2, also representing a total of seven statements (Nelson and Platnick, 1991: 360). However, statements A(BC), A(BD) and A(CD) are not independent over the entire data set and fractional weighting is required to adjust those values (see Table 9.9). After adjustment, tree (D(A(BC) still accommodates four statements but with different values (rounding factor of × 10 – the fractional weights have been rounded by a factor of 10 in this example to allow the programs to deal with whole numbers): A(BC) × 27, D(AB) × 10, D(AC) xx 10 and D(BC) × 20, representing a total of 67 statements, while tree (A(D(BC) still accommodates four statements also with different values (rounding factor of × 10): A(BC) × 27, A(BD) × 7, A(CD) × 7 and D(BC) × 20, representing a total of 61 statements, six fewer than the alternative tree and not a most parsimonious representation of the data.

Matrix 13 has one additional BC(AD) character yielding an additional two statements (Table 9.10). In this case, tree (D(A(BC) accommodates the same four statements A(BC) × 3, D(AB) × 1, D(AC) × 2 and D(BC) × 2, representing a total of

Table 9.11 Matrix 15 (after Nelson and Platnick, 1991)

O	0	0	0	0	0	0	0
A	1	0	?	?	?	?	?
B	1	1	0	?	?	?	?
C	1	1	1	0	?	?	0
D	1	?	1	1	0	?	1
E	0	?	?	1	1	0	1
F	0	?	?	?	1	1	?
G	0	?	?	?	?	1	?

eight statements with uniform weighting, and A(BC) × 27, D(AB) × 10, D(AC) × 20 and D(BC) × 20, representing a total of 77 statements with fractional weighting. Tree (A(D(BC) differs in this example. It also accommodates four statements, A(BC) × 3, A(BD) × 1, A(CD) × 1 and D(BC) × 2, but this represents a total of only seven statements, one fewer than the competing tree and hence less parsimonious. (This holds for fractional weighting as well in this instance.) Three-item analysis seems generally more sensitive to the accumulation of new data (Nelson, 1996).

Example g

(Matrix 15 of Nelson and Platnick, 1991; reproduced in Farris, 1997: Fig. 6.) This example has been discussed in more detail elsewhere (Siebert and Williams, 1998). Three-item analysis is unaffected by fractional weighting. Nelson and Platnick (1991) presented matrix 15 as an example of a three-item analysis with actual missing data. Three-item analysis sees character 1 (Table 9.11) as support for A–D, hence yielding the (A(B(CD)))(E(FG)) cladogram. Conventional analysis sees character 1 as 'silent' support for E–G; silent in that it is considered an informative 'reversal'; three-item analysis draws support from the original primary homology conjecture, EFG(ABCD).

Example i

(Matrix 17 minus one character and Matrix 18 minus one suite of statements of Nelson and Platnick, 1991; reproduced in Farris, 1997: Fig. 5.) This example has been discussed in more detail elsewhere (Siebert and Williams, 1998). Three-item analysis with elimination of overly resolved nodes yields two minimal trees rather than the four reported by Nelson and Platnick. Neither of the two minimal trees contains the (DE) or (DF) grouping objected to by Farris (1997).

Example j

(Matrix 19 and 20 of Nelson and Platnick, 1991.) Matrix 19 is of a similar kind to those discussed in Nelson (1996) and needs no further elaboration here.

Harvey (1992)

Harvey presented a series of matrices (Harvey, 1992: 350, Fig. 4, his matrices A-I) designed to demonstrate differences between three-item and conventional analysis. Nelson (1992: 355) examined the matrices from a three-item perspective and concluded: '[o]ne fact is that virtually all information in his [Harvey's] trees is present in the three-item trees'; the 'virtually' referred to the fact that 'group ABC of data set D is the exception' (Nelson, 1992: 355). Re-analysis of Harvey's matrices A–I using fractional weighting yields similar results to those reported by Nelson with the exception of Harvey's matrix D. Analysis of this matrix with fractional weighting yields two trees of which the strict consensus is F(E(D(ABC), identical to conventional analysis. Thus, with respect to Harvey's matrices, one fact is that *all* information in his trees is present in the three-item trees.

Kluge (1994)

Kluge's third example (Kluge, 1994: Table 3, Fig. 3; reproduced and expanded in Farris, 1997: 135, his matrix Z – see also Kluge, 1994: 413) yields an identical tree to conventional analysis when fractional weights are applied, as does Farris' (1997: matrix Z) modification (for further details see Siebert and Williams (1998)).

Nelson (1996)

Re-analysis of Nelson's matrices using fractional weighting and minimal trees does not alter the conclusions he presented.

Goloboff (1997)

We can make no judgement on Goloboff's results, as he gives no indication as to whether fractional weighting was used in the analyses (Goloboff, 1997: 238, 245, Table 2).

Conclusions

With respect to the general understanding of primary homology, there is no substitute for acquiring a deeper understanding of anatomy, cell ultrastructure and morphology of all organisms, as well as a deeper understanding of molecular data (see, for instance, Naylor and Brown, 1997). If nothing else, such studies tell us something new about the biology of the organisms of choice. With respect to the representation of data, greater attention needs to be given to the *actual* information content of our observations rather than what particular columns or combination of columns in any matrix yield upon analysis (i.e. Pleijel, 1995; Hawkins *et al.*, 1997; Chapters 2 and 3, this volume). Limited progress will result from supporting one or another view with respect to any particular implementation simply because it is somehow more pleasing, or confers some overall agreement among contemporaries.

Three-item analysis avoids the formalisms usually associated with comparative data analysis. Examination of how and why three-item analysis works is limited to the decisive matrices explored by Nelson (1996) (matrix 19 from Nelson and Platnick (1991) is another example of this kind). We have suggested elsewhere (Siebert and Williams, 1998) that these matrices are outside the scope of conventional analysis *because* of character optimization. This is because the kind of optimization involved implies a particular model of character evolution – a 'model of character evolution that requires synapomorphy to have unique origin (optimized as 1 at a node, with distal 0s as reversal, or vice versa)' (Nelson, 1992: 360). Rejecting three-item analysis for its potential ability to perform beyond the limits of conventional analysis is of little significance: the model underlying conventional analysis cannot 'explain' some data. Conventional parsimony analysis does imply a particular model of character evolution (a linear transformation of one state into another) – no model can be ever be general enough. In this we agree with Carpenter (1994: 218): 'Model-dependent approaches to phylogenetic inference inevitably discard evidence, because the most general models do not apply in specific cases and the highly specific models cannot be general'. In any case, a preoccupation with representations and formalisms of one particular kind hardly does justice to data carefully gathered.

Cladistics had its first wave of success as many long-standing problems were successively and efficiently solved and, once recognized for what they are, paraphyletic taxa disappear from our classifications. Today things are different. While some continue to dwell on the potential of paraphyly, the greatest challenge facing systematists is what to do with the wealth of data that stubbornly show varying degrees of conflict, even in the face of rigorous analysis. That wealth is recorded in our observations and our understanding of what those observations might tell us of the world. Perhaps, after all, it is the nature of that conflict we should be studying, which may be captured by a greater understanding of the nature of 'characters' (Nelson, 1978: 344) rather than the behaviour of various permutations of data matrices.

References

Brady, R.H. (1994a) Pattern description, process explanation, and the history of morphological sciences, in Grande, L. and Rieppel, O. (eds) *Interpreting the Hierarchy of Nature. From Systematic Patterns to Evolutionary Process Theories*, San Diego, CA: Academic Press, pp. 7–31.

Brady, R.H. (1994b). Explanation, description, and the meaning of 'transformation' in taxonomic evidence, in Scotland, R.W., Siebert, D.J. and Williams, D.M. (eds) *Models in Phylogeny Reconstruction*, Oxford: Clarendon Press, pp. 11–29.

Brower, A.V.Z. and Schawaroch, V. (1996) Three steps of homology assessment, *Cladistics*, 12, 265–272.

Carpenter, J.M. (1994) Successive weighting, reliability and evidence, *Cladistics*, 10, 215–220.

Coddington, J. and Scharff, N. (1994) Problems with zero-length branches, *Cladistics*, 10, 415–423.

De Laet, J. (1997). *A Reconsideration of Three-item Analysis, the Use of Implied Weights in Cladistics, and a Practical Application in Gentianaceae*, Proefschrift voorgedragen tot het behalen van de graad van Doctor in de Wetenschappen, Lueven.

Deleporte, P. (1996) Three-taxon statements and phylogeny construction, *Cladistics*, 12, 273–289.

De Pinna, M.C.C. (1991) Concepts and tests of homology in the cladistic paradigm, *Cladistics*, 7, 367–394.

De Pinna, M.C.C. (1996) Comparative biology and systematics: some controversies in retrospective, *Journal of Comparative Biology*, 1, 3–15.

Eldredge, N. and Cracraft, J. (eds) (1980) *Phylogenetic Patterns and the Evolutionary Process: Method and Theory in Comparative Biology*, New York: Columbia University Press.

Farris, J.S. (1997) Cycles, *Cladistics*, **13**, 131–144.

Farris, J.S., Kallersjo, M., Albert, V.A., Allard, M., Anderberg, A., Bowditch, B., Bult, C., Carpenter, J.M., Crowe, T.M., De Laet, J. *et al.*, (1995) Explanation, *Cladistics*, **11**, 211–218.

Farris, J.S., Kluge, A.G. and Eckhart, M.J. (1970) A numerical approach to phylogenetic systematics, *Systematic Zoology*, **19**, 172–189.

Felsenstein, J. (1978) The number of evolutionary trees, *Systematic Zoology*, **27**, 27–33.

Fitch, W.M. (1971) Toward defining the course of evolution: minimum change for a specific tree topology, *Systematic Zoology*, **20**, 406–416.

Fitch, W.M. (1975) The relationship between Prim networks and trees of maximum parsimony, in Estabrook, G.F. (ed) *Proceedings of the Eighth International Conference on Numerical Taxonomy*, San Francisco, CA: W.H. Freeman, pp. 189–230.

Gardiner, B.G., Janvier, P., Patterson, C., Forey, P.L., Greenwood, P.H., Miles, R.S. and Jefferies, R.P.S. (1979) The salmon, the lungfish and the cow; a reply, *Nature*, **277**, 175–176.

Goloboff, P.A. (1993) *NONA, ver. 1.51. Program and Documentation*, published by the author.

Goloboff, P.A. (1997) Self-weighted optimization: tree searches and character state reconstructions under implied transformation costs, *Cladistics*, **13**, 225–245.

Grande, L. (1994) Repeating patterns in nature, predictability, and "impact" in science, in Grande, L. and Rieppel, O. (eds) *Interpreting the Hierarchy of Nature. From Systematic Patterns to Evolutionary Process Theories*, San Diego, CA: Academic Press, pp. 61–84.

Harvey, A.W. (1992) Three-taxon statements: more precisely, an abuse of parsimony? *Cladistics*, **8**, 345–354.

Hawkins, J.A., Hughes, C.E. and Scotland, R.W. (1997) Primary homology assessment, characters and character states, *Cladistics*, **13**, 275–283.

Hennig, W. (1966) *Phylogenetic Systematics*, Urbana, IL: University of Illinois Press.

Hennipman, E. and Roos, M.C. (1983) Phylogenetic systematics of the Polypodiaceae, *Verhandlungen des Naturwissenschaftlichen Vereins in Hamburg*, **26**, 321–342.

Hull, D.L. (1973) *Darwin and his Critics: the Reception of Darwin's Theory of Evolution by the Scientific Community*, Cambridge, MA: Harvard University Press.

Huxley, T.H. (1888) The gentians: notes and queries, *Journal of the Linnean Society, Botany*, **24**, 101–124.

Kitching, I.J., Forey, P., Humphries, C.J. and Williams, D.M. (1998) *Cladistics: the Theory and Practice of Parsimony Analysis*, Oxford: Oxford University Press.

Kluge, A.G. (1993) Three-taxon transformation in phylogenetic inference: ambiguity and distortion as regards explanatory power, *Cladistics*, **9**, 246–259.

Kluge, A.G. (1994) Moving targets and shell games, *Cladistics*, **10**, 403–413.

Lloyd, D.G. and Calder, V.L. (1991) Multi-residue gaps, a class of molecular characters with exceptional reliability for phylogenetic analysis, *Journal of Evolutionary Biology*, **4**, 9–21.

Lyons-Weiler, J., Hoelzer, G.A. and Tausch, R.J. (1996) Relative apparent synapomorphy (RASA) I: The statistical measurement of phylogenetic signal, *Molecular Biology and Evolution*, **13**, 749–757.

Maddison, D.R., Swofford, D.L. and Maddison, W.P. (1997) Nexus: an extensible file format for systematic information, *Systematic Biology*, **46**, 590–621.

Mickevich, M.F. and Weller, S.J. (1990) Evolutionary character analysis: tracing character change on a cladogram, *Cladistics*, **6**, 137–170.

Morrone, J.J. and Carpenter, J.M. (1994) In search of a method for cladistic biogeography: an empirical comparison of component analysis, Brooks parsimony analysis, and three-area statements, *Cladistics*, **10**, 99–153.

Naylor, G.J.P. and Brown, W.M. (1997) Structural biology and phylogenetic estimation, *Nature*, **388**, 527–528.

Nelson, G.J. (1978) Ontogeny, phylogeny, paleontology, and the biogenetic law, *Systematic Zoology*, **27**, 324–345.

Nelson, G.J. (1979) Cladistic analysis and synthesis: principles and definitions, with a historical note on Adanson's *Families des Plantes* (1763–1764), *Systematic Zoology*, **28**, 1–21.

Nelson, G.J. (1992) Reply to Harvey, *Cladistics*, **8**, 355–360.

Nelson, G.J. (1993) Reply, *Cladistics*, **9**, 261–265.

Nelson, G.J. (1994) Homology and systematics, in Hall, B.K. (ed.) *Homology: the Hierarchical Basis of Comparative Biology*, San Diego, CA: Academic Press, pp. 101–149.

Nelson, G.J. (1996), *Nullius in verba*, published by the author, New York (reprinted in *Journal of Comparative Biology*, **1**, 141–152).

Nelson, G.J. and Ladiges, P.Y. (1992a) Three-area statements: standard assumptions for biogeographic analysis, *Systematic Zoology*, **40**, 470–485.

Nelson, G.J. and Ladiges, P.Y. (1992b) Information content and fractional weight of three-taxon statements, *Systematic Biology*, **41**, 490–494.

Nelson, G.J. and Ladiges, P.Y. (1993) Missing data and three-item analysis, *Cladistics*, **9**, 111–113.

Nelson, G.J. and Ladiges, P.Y. (1994a) Three-item consensus: empirical test of fractional weighting, in Scotland, R.W., Siebert, D.J. and Williams, D.M. (eds) *Models in Phylogeny Reconstruction*, Oxford: Clarendon Press, pp. 193–209.

Nelson, G.J. and Ladiges, P.Y. (1994b) *TAX: MSDOS Program for Cladistics*, New York and Melbourne.

Nelson, G.J. and Ladiges, P.Y. (1994c) *TAS: Three Area Statement Analysis*, New York and Melbourne.

Nelson, G.J. and Ladiges, P.Y. (1996) Paralogy in cladistic biogeography and analysis of paralogy-free subtrees, *American Museum Novitates*, **3167**, 1–58.

Nelson, G.J. and Platnick, N.I. (1981). *Systematics and Biogeography: Cladistics and Vicariance*, New York: Columbia University Press.

Nelson, G.J. and Platnick, N.I. (1984) Systematics and evolution, in Ho, M.W. and Saunders, P. (eds) *Beyond Neodarwinism*, London: Academic Press, pp. 143–158.

Nelson, G.J. and Platnick, N.I. (1991) Three-taxon statements: a more precise use of parsimony? *Cladistics*, **7**, 351–366.

Nixon, K.C. (1996) Paleobotany in cladistics and cladistics in paleobotany: enlightenment and uncertainty, *Review of Palaeobotany and Palynology*, **90**, 361–373.

Nixon, K. and Carpenter, J.M. (1996) On consensus, collapsibility, and clade concordance, *Cladistics*, **12**, 305–321.

Page, R.D.M. (1990) Component analysis: a valiant failure? *Cladistics*, **6**, 119–136.

Page, R.D.M. (1992) Comments on the information content of classifications, *Cladistics*, **8**, 87–95.

Patterson, C. (1981) Methods in paleobiogeography, in Nelson, G.J. and Rosen, D.E. (eds) *Vicariance Biogeography: a Critique* New York: Columbia University Press, pp. 446–489.

Patterson, C. (1982) Morphological characters and homology, in Joysey, K.A. and Friday, A.E. (eds) *Problems of Phylogeny Reconstruction*, London: Academic Press, pp. 21–74.

Patterson, C. and Johnson, G.D. (1995) The intermuscular bones and ligaments of teleostean fishes, *Smithsonian Contributions in Zoology*, **559**, 1–85.

Platnick, N.I. (1977) Cladograms, phylogenetic trees, and hypothesis testing, *Systematic Zoology*, **26**, 438–442.

Platnick, N.I. (1979) Philosophy and the transformation of cladistics, *Systematic Zoology*, 28, 537–546.

Platnick, N.I. (1987) An empirical comparison of microcomputer parsimony programs, *Cladistics*, 3, 121–144.

Platnick, N.I. (1989a) An empirical comparison of microcomputer parsimony programs, II, *Cladistics*, 5, 145–161.

Platnick, N.I. (1989b) Cladistic and phylogenetic analysis today, in Fernholm, B., Bremer, K, and Jörnvall, H. (eds) *The Hierarchy of Life*, Amsterdam: Excerpta Medica, pp. 17–24.

Platnick, N.I. (1993) Character optimization and weighting: differences between the standard and three-taxon approaches to phylogenetic inference, *Cladistics*, 9, 267–272.

Platnick, N.I., Humphries C.J., Nelson, G.J. and Williams, D.M. (1996) Is Farris optimization perfect? *Cladistics*, 12, 243–252.

Pleijel, F. (1995) On character coding for phylogeny reconstruction, *Cladistics*, 11, 309–315.

Purvis, A. (1995) A modification to Baum and Ragan's method for combining phylogenetic trees, *Systematic Biology*, 44, 251–255.

Ronquist, F. (1996) Matrix representation of trees, redundancy, and weighting, *Systematic Biology*, 45, 247–253.

Siebert, D.J. and Williams, D.M. (1997) Book review [*Nullius in Verba*], *Biological Journal of the Linnean Society*, 60, 145–146.

Siebert, D.J. and Williams, D.M. (1998) Recycled, *Cladistics*, 14, 339–347.

Udovicic, F., McFadden, G.I. and Ladiges, P.Y. (1995) Phylogeny of *Eucalyptus* and *Angophora* based on 5s rDNA spacer sequence data, *Molecular Phylogenetics and Evolution*, 4, 247–256.

Wägele, J.W. (1994) Review of methodological problems of 'computer cladistics' exemplified with a case study on isopod phylogeny (Crustacea: Isopoda), *Zeitschrift für Zoologische Systematik und Evolutionsforschung*, 32, 81–107.

Wheeler, W. (1990) Combinatorial weights in phylogenetic analysis: a statistical parsimony procedure, *Cladistics*, 6, 269–275.

Wilkinson, M. (1994a) Common cladistic information and its consensus representation: reduced Adams and reduced cladistic consensus trees and profiles, *Systematic Biology*, 43, 343–368.

Wilkinson, M. (1994b) Three-taxon statements: when is a parsimony analysis also a clique analysis? *Cladistics*, 10, 221–223.

Wilkinson, M. (1995) Arbitrary resolutions, missing entries, and the problem of zero-length branches in parsimony analysis, *Systematic Biology*, 44, 108–111.

Williams, D.M. (1996a) Characters and cladograms, *Taxon*, 45, 275–283.

Williams, D.M. (1996b) Fossil species of the diatom genus *Tetracyclus* (Bacillariophyta, 'ellipticus' species group): morphology, interrelationships and the relevance of ontogeny, *Philosophical Transactions of the Royal Society, London*, B351, 1759–1782.

Index

Systematics Association Publications

1. Bibliography of key works for the identification of the British fauna and flora, 3rd edition (1967)[†]
Edited by G.J. Kerrish, R.D. Meikle and N. Tebble
2. Function and taxonomic importance (1959)[†]
Edited by A.J. Cain
3. The species concept in palaeontology (1956)[†]
Edited by P.C. Sylvester-Bradley
4. Taxonomy and geography (1962)[†]
Edited by D. Nichols
5. Speciation in the sea (1963)[†]
Edited by J.P. Harding and N. Tebble
6. Phenetic and phylogenetic classification (1964)[†]
Edited by V.H. Heywood and J. McNeill
7. Aspects of Tethyan biogeography (1967)[†]
Edited by C.G. Adams and D.V. Ager
8. The soil ecosystem (1969)[†]
Edited by H. Sheals
9. Organisms and continents through time (1973)[†]
Edited by N.F. Hughes
10. Cladistics: a practical course in systematics (1992)[*]
P.L. Forey, C.J. Humphries, I.J. Kitching, R.W. Scotland, D.J. Siebert and D.M. Williams
11. Cladistics: the theory and practice of parsimony analysis (2nd edition) (1998)[*]
I.J. Kitching, P.L. Forey, C.J. Humphries and D.M. Williams

[*]Published by Oxford University Press for the Systematics Association
[†] Published by the Association (out of print)

Systematics Association Special Volumes

1. The new systematics (1940)
Edited by J.S. Huxley (reprinted 1971)
2. Chemotaxonomy and serotaxonomy (1968)[*]
Edited by J.G. Hawkes
3. Data processing in biology and geology (1971)[*]
Edited by J.L. Cutbill
4. Scanning electron microscopy (1971)[*]
Edited by V.H. Heywood
5. Taxonomy and ecology (1973)[*]
Edited by V.H. Heywood
6. The changing flora and fauna of Britain (1974)[*]
Edited by D.L. Hawksworth

7. Biological identification with computers (1975)*
Edited by R.J. Pankhurst
8. Lichenology: progress and problems (1976)*
Edited by D.H. Brown, D.L. Hawksworth and R.H. Bailey
9. Key works to the fauna and flora of the British Isles and northwestern Europe, 4th edition (1978)*
Edited by G.J. Kerrich, D.L. Hawksworth and R.W. Sims
10. Modern approaches to the taxonomy of red and brown algae (1978)
Edited by D.E.G. Irvine and J.H. Price
11. Biology and systematics of colonial organisms (1979)*
Edited by G. Larwood and B.R. Rosen
12. The origin of major invertebrate groups (1979)*
Edited by M.R. House
13. Advances in bryozoology (1979)*
Edited by G.P. Larwood and M.B. Abbott
14. Bryophyte systematics (1979)*
Edited by G.C.S. Clarke and J.G. Duckett
15. The terrestrial environment and the origin of land vertebrates (1980)
Edited by A.L. Pachen
16. Chemosystematics: principles and practice (1980)*
Edited by F.A. Bisby, J.G. Vaughan and C.A. Wright
17. The shore environment: methods and ecosystems (2 volumes) (1980)*
Edited by J.H. Price, D.E.G. Irvine and W.F. Farnham
18. The Ammonoidea (1981)*
Edited by M.R. House and J.R. Senior
19. Biosystematics of social insects (1981)*
Edited by P.E. House and J.-L. Clement
20. Genome evolution (1982)*
Edited by G.A. Dover and R.B. Flavell
21. Problems of phylogenetic reconstruction (1982)*
Edited by K.A. Joysey and A.E. Friday
22. Concepts in nematode systematics (1983)*
Edited by A.R. Stone, H.M. Platt and L.F. Khalil
23. Evolution, time and space: the emergence of the biosphere (1983)*
Edited by R.W. Sims, J.H. Price and P.E.S. Whalley
24. Protein polymorphism: adaptive and taxonomic significance (1983)*
Edited by G.S. Oxford and D. Rollinson
25. Current concepts in plant taxonomy (1983)*
Edited by V.H. Heywood and D.M. Moore
26. Databases in systematics (1984)*
Edited by R. Allkin and F.A. Bisby
27. Systematics of the green algae (1984)*
Edited by D.E.G. Irvine and D.M. John
28. The origins and relationships of lower invertebrates (1985)‡
Edited by S. Conway Morris, J.D. George, R. Gibson and H.M. Platt
29. Infraspecific classification of wild and cultivated plants (1986)‡
Edited by B.T. Styles

30. Biomineralization in lower plants and animals (1986)[‡]
Edited by B.S.C. Leadbeater and R. Riding
31. Systematic and taxonomic approaches in palaeobotany (1986)[‡]
Edited by R.A. Spicer and B.A. Thomas
32. Coevolution and systematics (1986)[‡]
Edited by A.R. Stone and D.L. Hawksworth
33. Key works to the fauna and flora of the British isles and northwestern Europe, 5th edition (1988)[‡]
Edited by R.W. Sims, P. Freeman and D.L. Hawksworth
34. Extinction and survival in the fossil record (1988)[‡]
Edited by G.P. Larwood
35. The phylogeny and classification of the tetrapods (2 volumes) (1988)[‡]
Edited by M.J. Benton
36. Prospects in systematics (1988)[‡]
Edited by D.L. Hawksworth
37. Biosystematics of haematophagous insects (1988)[‡]
Edited by M.W. Service
38. The chromophyte algae: problems and perspective (1989)[‡]
Edited by J.C. Green, B.S.C. Leadbeater and W. L. Diver
39. Electrophoretic studies on agricultural pests (1989)[‡]
Edited by H.D. Loxdale and J. den Hollander
40. Evolution, systematics, and fossil history of the Hamamelidae (2 volumes) (1989)[‡]
Edited by P.R. Crane and S. Blackmore
41. Scanning electron microscopy in taxonomy and functional morphology (1990)[‡]
Edited by D. Claugher
42. Major evolutionary radiations (1990)[‡]
Edited by P.D. Taylor and G.P. Larwood
43. Tropical lichens: their systematics, conservation and ecology (1991)[‡]
Edited by G.J. Galloway
44. Pollen and spores: patterns of diversification (1991)[‡]
Edited by S. Blackmore and S.H. Barnes
45. The biology of free-living heterotrophic flagellates (1991)[‡]
Edited by D.J. Patterson and J. Larsen
46. Plant-animal interactions in the marine benthos (1992)[‡]
Edited by D.M. John, S.J. Hawkins and J.H. Price
47. The Ammonoidea: environment, ecology and evolutionary change (1993)[‡]
Edited by M.R. House
48. Designs for a global plant species information system (1993)[‡]
Edited by F.A. Bisby, G.F. Russell and R.J. Pankhurst
49. Plant galls: organisms, interactions, populations (1994)[‡]
Edited by M.A.J. Williams
50. Systematics and conservation evaluation (1994)[‡]
Edited by P.L. Forey, C.J. Humphries and R.I. Vane-Wright
51. The Haptophyte algae (1994)[‡]
Edited by J.C. Green and B.S.C. Leadbeater

52. Models in phylogeny reconstruction (1994)‡
Edited by R. Scotland, D.I. Siebert, and D.M. Williams
53. The ecology of agricultural pests: biochemical approaches (1996)**
Edited by W.O.C. Symondson and J.E. Liddell
54. Species: the units of diversity (1997)**
Edited by M.F. Claridge, H.A. Dawah and M.R. Wilson
55. Arthropod relationships (1998)**
Edited by R.A. Fortey and R.H. Thomas
56. Evolutionary relationships among Protozoa (1998)**
Edited by G.H. Coombs, K. Vickerman, M.A. Sleigh and A. Warren
57. Molecular systematics and plant evolution
Edited by P.M. Hollingsworth, R.M. Bateman and R.J. Gornall
58. Homology and Systematics: Coding characters for phylogenetic analysis
Edited by R. Scotland and R.T. Pennington

* Published by Academic Press for the Systematics Association
† Published by the Palaeontological Association in conjunction with the Systematics
   Association
‡ Published by the Oxford University Press for the Systematics Association
**Published by Chapman and Hall for the Systematics Association

Milton Keynes UK
Ingram Content Group UK Ltd.
UKHW051952071024
449327UK00026B/2274